Model-Assisted Bayesian Designs for Dose Finding and Optimization
Methods and Applications

Bayesian adaptive designs provide a critical approach to improve the efficiency and success of drug development that has been embraced by the US Food and Drug Administration (FDA). This is particularly important for early phase trials as they form the basis for the development and success of subsequent phase II and III trials.

The objective of this book is to describe the state-of-the-art model-assisted designs to facilitate and accelerate the use of novel adaptive designs for early phase clinical trials. Model-assisted designs possess avant-garde features where superiority meets simplicity. Model-assisted designs enjoy exceptional performance comparable to more complicated model-based adaptive designs, yet their decision rules often can be pre-tabulated and included in the protocol—making implementation as simple as conventional algorithm-based designs. An example is the Bayesian optimal interval (BOIN) design, the first dose-finding design to receive the fit-for-purpose designation from the FDA. This designation underscores the regulatory agency's support of the use of the novel adaptive design to improve drug development.

Features

- Represents the first book to provide comprehensive coverage of model-assisted designs for various types of dose-finding and optimization clinical trials
- Describes the up-to-date theory and practice for model-assisted designs
- Presents many practical challenges, issues, and solutions arising from early-phase clinical trials
- Illustrates with many real trial applications
- Offers numerous tips and guidance on designing dose finding and optimization trials
- Provides step-by-step illustrations of using software to design trials
- Develops a companion website (www.trialdesign.org) to provide freely available, easy-to-use software to assist learning and implementing model-assisted designs

Written by internationally recognized research leaders who pioneered model-assisted designs from the University of Texas MD Anderson Cancer Center, this book shows how model-assisted designs can greatly improve the efficiency and simplify the design, conduct, and optimization of early-phase dose-finding trials. It should therefore be a very useful practical reference for biostatisticians, clinicians working in clinical trials, and drug regulatory professionals, as well as graduate students of biostatistics. Novel model-assisted designs showcase the new KISS principle: Keep it simple and smart!

Chapman & Hall/CRC Biostatistics Series

Series Editors
Shein-Chung Chow, Duke University School of Medicine, USA
Byron Jones, Novartis Pharma AG, Switzerland
Jen-pei Liu, National Taiwan University, Taiwan
Karl E. Peace, Georgia Southern University, USA
Bruce W. Turnbull, Cornell University, USA

Recently Published Titles

Statistical Thinking in Clinical Trials
Michael A. Proschan

Simultaneous Global New Drug Development
Multi-Regional Clinical Trials after ICH E17
Edited by Gang Li, Bruce Binkowitz, William Wang, Hui Quan, and Josh Chen

Quantitative Methodologies and Process for Safety Monitoring and Ongoing Benefit Risk Evaluation
Edited by William Wang, Melvin Munsaka, James Buchanan and Judy Li

Statistical Methods for Mediation, Confounding and Moderation Analysis Using R and SAS
Qingzhao Yu and Bin Li

Hybrid Frequentist/Bayesian Power and Bayesian Power in Planning Clinical Trials
Andrew P. Grieve

Advanced Statistics in Regulatory Critical Clinical Initiatives
Edited By Wei Zhang, Fangrong Yan, Feng Chen, Shein-Chung Chow

Medical Statistics for Cancer Studies
Trevor F. Cox

Real World Evidence in a Patient-Centric Digital Era
Edited by Kelly H. Zou, Lobna A. Salem, Amrit Ray

Data Science, AI, and Machine Learning in Pharma
Harry Yang

Model-Assisted Bayesian Designs for Dose Finding and Optimization
Methods and Applications
Ying Yuan, Ruitao Lin, and J. Jack Lee

For more information about this series, please visit: https://www.routledge.com/ Chapman--Hall-CRC-Biostatistics-Series/book-series/CHBIOSTATIS

Model-Assisted Bayesian Designs for Dose Finding and Optimization

Methods and Applications

Ying Yuan

The University of Texas MD Anderson Cancer Center, USA

Ruitao Lin

The University of Texas MD Anderson Cancer Center, USA

J. Jack Lee

The University of Texas MD Anderson Cancer Center, USA

CRC Press

Taylor & Francis Group

Boca Raton London New York

CRC Press is an imprint of the
Taylor & Francis Group, an **informa** business

A CHAPMAN & HALL BOOK

First edition published 2023
by CRC Press
6000 Broken Sound Parkway NW, Suite 300, Boca Raton, FL 33487-2742

and by CRC Press
4 Park Square, Milton Park, Abingdon, Oxon, OX14 4RN

© 2023 Taylor & Francis Group, LLC

CRC Press is an imprint of Taylor & Francis Group, LLC

Library of Congress Cataloging-in-Publication Data

Names: Yuan, Ying (Professor of biostatistics), author. | Lin, Ruitao,
   author. | Lee, J. Jack, author.
Title: Model-assisted Bayesian designs for dose finding and optimization :
   methods and applications / Ying Yuan, Ruitao Lin, J. Jack Lee.
Other titles: Chapman & Hall/CRC biostatistics series.
Description: First edition. | Boca Raton : C&Hall/CRC Press, 2023. |
   Series: Chapman & Hall/CRC biostatistics series | Includes
   bibliographical references and index.
Identifiers: LCCN 2022017949 (print) | LCCN 2022017950 (ebook) | ISBN
   9780367146245 (hardback) | ISBN 9781032357126 (paperback) | ISBN
   9780429052781 (ebook)
Subjects: MESH: Drug Development--methods | Drug Design--methods | Drug
   Dosage Calculations | Models, Statistical | Bayes Theorem | Clinical
   Trials, Phase I as Topic
Classification: LCC RM301.25 (print) | LCC RM301.25 (ebook) | NLM QV 745
   | DDC 615.1/9--dc23/eng/20220801
LC record available at https://lccn.loc.gov/2022017949
LC ebook record available at https://lccn.loc.gov/2022017950

ISBN: 978-0-367-14624-5 (hbk)
ISBN: 978-1-032-35712-6 (pbk)
ISBN: 978-0-429-05278-1 (ebk)

DOI: 10.1201/9780429052781

Typeset in CMR10
by KnowledgeWorks Global Ltd.

Publisher's note: This book has been prepared from camera-ready copy provided by the authors.

To my wife, Suyu, and daughter, Selina.

Ying Yuan

To the memory of my mother, a brave cancer fighter.

Ruitao Lin

To my wife, Vei-Vei, and children, Joseph and Hope, for their love and support.

J. Jack Lee

Contents

Preface

Despite extensive efforts, the drug development enterprise is challenged by prohibitively high costs and slow progress. To address these pressing issues, one important approach embraced by the US Food and Drug Administration (FDA) is the use of novel adaptive designs. This is particularly important for early phase clinical trials, which serve as the gatekeepers to identifying promising drugs and optimizing treatment regimens (e.g., dose) before launching expensive phase III trials. A major barrier to the use of these tools is that most novel adaptive designs are difficult to understand, requiring complicated statistical modeling, complex computation, and expensive infrastructure for implementation.

The objective of this book is to systematically introduce and review a class of novel adaptive designs, known as model-assisted designs, to remove this barrier while increasing the use of novel adaptive designs for early phase clinical trials. Model-assisted designs enjoy superior performance comparable to more complicated model-based adaptive designs, yet their decision rule often can be pre-tabulated and included in the protocol—making implementation as simple as conventional algorithm-based designs. Thus, these novel model-assisted design features create a space where superiority meets simplicity. They adhere to the new KISS principle: Keep it simple and smart! A typical example is the Bayesian optimal interval (BOIN) design, the first maximum-tolerated dose (MTD)-finding design to receive the fit-for-purpose designation from the FDA as a tool for drug development. This designation underscores the regulatory agency's support for the use of novel adaptive design to improve drug development.

This book is intended for biostatisticians working on the design and analysis of clinical trials, as well as clinical trialists with an interest in learning adaptive designs. One unique feature of this book is that, besides statistical methodology, it also provides a detailed description of each design's software and step-by-step guidance for clinical trial design and conduct using real-world examples. A great amount of effort has been devoted to developing the companion website www.trialdesign.org to provide freely available, easy-to-use software, along with extensive help and documentation, to assist readers in learning and implementing the designs in this book.

Our hope is that this book and the accompanying software will provide a practical toolkit for practitioners to implement novel adaptive designs for their own trials. Our overarching goal is to benefit patients by accelerating drug development and clinical research.

It is almost impossible to thank all of the people who have contributed in various ways to making this book a reality. We are extremely grateful to all of our students, postdocs, and colleagues who helped develop the novel methodology and software in this book—in particular, Suyu Liu, Ph.D.; Peter F. Thall, Ph.D.; Kenneth R. Hess, Ph.D.; Susan G. Hilsenbeck, Ph.D.; Mark R. Gilbert, M.D.; Jing Wu, M.D.; Guosheng Yin, Ph.D.; Sumithra J. Mandrekar, Ph.D.; Yanhong Zhou, Ph.D.; Heng Zhou, Ph.D.; Fangrong Yan, Ph.D.; Haitao Pan, Ph.D.; Rongji Mu, Ph.D.; Kai Chen, M.S.; Ying-Wei Kuo, M.S.; and Nan Chen, Ph.D. We would also like to express our sincere gratitude to our industrial or regulatory collaborators including Daniel Li, Ph.D.; Lei Nie, Ph.D.; Sammi Tang, Ph.D.; Li Wang, Ph.D.; and Zhiying Pan, Ph.D. for their insightful discussions and suggestions. We thank Jessica Swann for her detailed reading and editorial assistance, and Evan Kwiatkowski, Grace Nie, Peng Yang, Feng Tian, Xiaohan Chi, Jingyi Zhang, Mengyi Lu, and other students for proofreading.

Ying Yuan
Ruitao Lin
J. Jack Lee

Author Biographies

Ying Yuan, Ph.D., is Bettyann Asche Murray Distinguished Professor in Biostatistics and Deputy Chair at the Department of Biostatistics at the University of Texas MD Anderson Cancer Center. He has published over 100 statistical methodology papers on innovative Bayesian adaptive designs, including early phase trials, seamless trials, biomarker-guided trials, and basket and platform trials. The designs and software developed by Dr. Yuan's and Dr. J. Jack Lee's team (www.trialdesign.org) have been widely used in medical research institutes and pharmaceutical companies. The BOIN design developed by Dr. Yuan's team is the first oncology dose-finding design designated as a fit-for-purpose drug development tool by FDA. Dr. Yuan is an elected Fellow of the American Statistical Association, and is a co-author of the book *Bayesian Designs for Phase I-II Clinical Trials* published by Chapman & Hall/CRC Press.

Ruitao Lin, Ph.D., is an Assistant Professor in the Department of Biostatistics at the University of Texas MD Anderson Cancer Center. Motivated by the unmet need for the development of precision medicine, Dr. Lin has developed many innovative statistical designs to increase trial efficiency, optimize healthcare decisions, and expedite drug development. He made substantial contributions to generalize model-assisted designs, including BOIN, to handle combination trials, late-onset toxicity, and dose optimization. Dr. Lin has published over 40 papers in top statistical and medical journals. He currently is an Associate Editor of *Biometrical Journal, Pharmaceutical Statistics*, and *Contemporary Clinical Trials*.

J. Jack Lee, Ph.D., is a Professor of Biostatistics, Kenedy Foundation Chair in Cancer Research, and Associate Vice President in Quantitative Sciences at the University of Texas MD Anderson Cancer Center. He is an expert on the design and analysis of Bayesian adaptive designs, platform trials, basket trials, umbrella trials, master protocols, statistical computation/graphics, drug combination studies, and biomarkers identification and validation. Dr. Lee has also been actively participating in basic, translational, and clinical cancer research in chemoprevention, immuno-oncology, and precision oncology. He is an elected Fellow of the American Statistical Association, the Society for Clinical Trials, and the American Association for the Advancement of Science. He is Statistical Editor of *Cancer Prevention Research* and serves on the Statistical Editorial Board of *Journal of the National Cancer Institute*. He has over 500 publications and is a co-author of the book *Bayesian Adaptive Methods for Clinical Trials* published by Chapman & Hall/CRC Press.

1

Bayesian Statistics and Adaptive Designs

1.1 Basics of Bayesian statistics

1.1.1 Bayes' theorem

Clinical trials are prospective studies to evaluate the health-related biomedical effect of one or more interventions in human subjects. An intervention could be a drug, medical device, or biologic such as a vaccine or gene therapy. Through clinical trials, data D are collected from study participants to inform the parameters of interest, θ, such as the treatment effect or the safety of the intervention. There are two major paradigms to make statistical inference on θ: frequentist and Bayesian. The frequentist paradigm assumes that θ is a fixed unknown quantity and D are random. The inference of θ is made by modeling the distribution of D, given θ, known as likelihood $L(D|\theta)$. The most common inferential method is the maximum likelihood method.

In contrast, the Bayesian paradigm assumes that D are fixed, as they have been already observed and do not change, and the unknown parameter θ is random. In the Bayesian framework, the unknown parameter has a distribution, which is used to quantify its uncertainty. Bayesian inference is based on the posterior distribution of θ given D. This is done through Bayes' theorem, which describes the conditional probability of an event A given an event B, formally:

$$\Pr(A \mid B) = \frac{\Pr(A \cap B)}{\Pr(B)} = \frac{\Pr(A)\Pr(B \mid A)}{\Pr(B)}.$$

Here, $\Pr(A) \geq 0$ and $\Pr(B) > 0$ are the probabilities of observing the events A and B, respectively. $\Pr(A \mid B)$ is the conditional probability of A given B occurring, and $\Pr(A \cap B)$ is the probability of both A and B occurring. Applying Bayes' theorem, the *posterior distribution* of θ, $f(\theta \mid D)$, is given by

$$f(\theta \mid D) = \frac{f(\theta)L(D \mid \theta)}{f(D)},$$

where $f(\theta)$ is the prior distribution of θ, quantifying the prior knowledge on θ, and $f(D) = \int f(\theta)L(D \mid \theta)d\theta$ is the *marginal likelihood* of D, serving as a normalizing constant to ensure that the posterior distribution is a valid density function, such that $\int f(\theta \mid D)d\theta = 1$. As the normalizing constant

DOI: 10.1201/9780429052781-1

$f(D)$ does not depend on θ, the (unnormalized) posterior distribution can also be expressed as

$$f(\theta \mid D) \propto f(\theta)L(D \mid \theta). \qquad (1.1)$$

This simpler expression encapsulates the essence of Bayesian inference: the posterior distribution is proportional to the product of the prior distribution and the likelihood function. In other words, the posterior distribution is the outcome of synthesizing the information contained in the prior distribution and data.

Compared to frequentist approaches, the Bayesian paradigm has several advantages that make it particularly suitable for clinical trial design. First, as shown in (1.1), it provides a convenient and principled way to incorporate prior information, when available, into the current clinical trial through the specification of the prior distribution $f(\theta)$. The prior information may come from preclinical data, results of clinical trials of the drug in different patient populations, and expert opinions. When the prior distribution is appropriately specified, Bayesian methods improve the trial efficiency and inference accuracy; see examples below.

Second, the Bayesian paradigm provides a coherent mechanism that allows us to continuously update our knowledge on θ and accordingly make appropriate clinical decisions based on accumulating data. Due to ethical considerations, clinical trials are often conducted in a group sequential way—treating patients by cohorts. Treatment decisions for later cohorts are made based on the outcomes of early cohorts. For example, in phase I trials, if the data from treated patients (cohorts) show that the current dose is overly toxic, we should de-escalate the dose to treat the next cohort. In phase II trials, if the data from treated patients indicate that the experimental drug is most likely ineffective or too toxic, we may terminate the trial early for futility or toxicity. Let $D_1, \cdots, D_K$ denote the data observed from cohorts $1, \cdots, K$. Under the Bayesian paradigm, our knowledge on θ is updated as follows: starting from the prior knowledge of θ, i.e., $f(\theta)$, after observing D_1, we update our knowledge of θ and obtain the posterior knowledge of θ as $f(\theta \mid D_1) \propto f(\theta)L(D_1 \mid \theta)$, which is used to guide the treatment decision for cohort 2. As this posterior represents our prior knowledge of θ before observing D_2, it serves as a prior distribution before the next step. After observing D_2, we apply Bayes' theorem and update our knowledge of θ again as follows:

$$f(\theta \mid D_1, D_2) \propto f(\theta|D_1)L(D_2 \mid \theta) \propto f(\theta)L(D_1 \mid \theta)L(D_2 \mid \theta).$$

This posterior is then used to guide the treatment decision for cohort 3. By continuing this process, we can learn about θ successively and refine treatment decisions for patients as the trial progresses.

Third, Bayesian methods conform to the likelihood principle, which states that all information for making inference on the parameters is contained in the observed data only and not in the unobserved data. Fourth, Bayesian modeling can naturally and appropriately address all levels of uncertainty in obtaining the data.

Example 1.1 (Beta-binomial model)   Suppose that we observe that y out of n patients have responded favorably by taking a new drug, the observed data $D = (y, n)$ can be modeled by a binomial distribution, $y \sim \text{Binomial}(n, \theta)$, with $0 \le \theta \le 1$ denoting the response rate. The likelihood function of D is

$$L(D \mid \theta) = \binom{n}{y} \theta^y (1 - \theta)^{n-y}.$$

Let $f(\theta)$ be the prior of θ, which takes a beta distribution, $\theta \sim \text{Beta}(\alpha, \beta)$ with the density of

$$f(\theta) = \frac{\Gamma(\alpha + \beta)}{\Gamma(\alpha)\Gamma(\beta)} \theta^{\alpha-1}(1 - \theta)^{\beta-1},$$

where $\alpha > 0$ and $\beta > 0$ are hyperparameters. Applying Bayes' theorem, the posterior distribution of θ is given by

$$\begin{aligned}
f(\theta \mid D) &\propto f(\theta)L(D \mid \theta) \\
&\propto \theta^{\alpha-1}(1 - \theta)^{\beta-1}\theta^y(1 - \theta)^{n-y} \\
&\propto \theta^{\alpha+y-1}(1 - \theta)^{\beta+n-y-1}.
\end{aligned} \tag{1.2}$$

Noting that (1.2) is the kernel of the beta distribution, the posterior distribution of θ follows a beta distribution

$$\theta \mid D \sim \text{Beta}(\alpha + y, \beta + n - y).$$

Thus, α and β can be interpreted as the prior numbers of responders and nonresponders, respectively, and $(\alpha + \beta)$ is the prior effective sample size. To see the impact of the prior on the posterior distribution, we assume $(y, n) = (10, 25)$ and consider the following three beta priors: (a) $\theta \sim \text{Beta}(0.5, 0.5)$, which is a vague (or weakly informative) prior with the prior effective sample size equal to one patient; (b) $\theta \sim \text{Beta}(2, 6)$, a moderately informative prior containing the information corresponding to eight patients; and (c) $\theta \sim \text{Beta}(150, 300)$, a strongly informative prior containing the information corresponding to 450 patients. The resulting posterior distributions based on the three priors are $\text{Beta}(10.5, 15.5)$, $\text{Beta}(12, 21)$, and $\text{Beta}(160, 315)$, respectively. Figure 1.1 exhibits the prior, likelihood (unnormalized), and posterior densities of θ based on the Beta-binomial model. It shows that when the prior is vague, the likelihood and posterior functions are almost identical, and when the prior is extremely informative, as exemplified by the prior $\text{Beta}(150, 300)$, the prior would dominate the posterior distribution, and the observed data play an extremely limited role in the inference of θ.

Example 1.2 (Normal-normal model with a known variance)   Suppose that in a clinical study of n patients, the outcome y_i for the ith patient

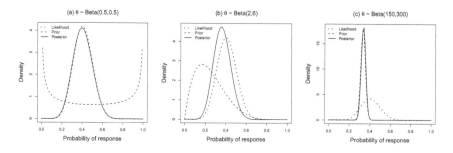

FIGURE 1.1: Prior, likelihood (unnormalized), and posterior densities for the probability of response θ, based on the Beta-binomial model with different beta priors.

follows a normal distribution $y_i \sim N(\mu, \sigma^2)$, $i = 1, \ldots, n$, where the mean μ is unknown and the variance σ^2 is known. Based on the data from n patients, $D = \{y_1, \ldots, y_n\}$, the likelihood function is given by

$$L(D \mid \mu) = \prod_{i=1}^{n} f(y_i \mid \mu, \sigma^2) = \frac{1}{\sigma^n (2\pi)^{n/2}} \exp\left\{ -\frac{1}{2\sigma^2} \sum_{i=1}^{n} (y_i - \mu)^2 \right\},$$

where $f(y_i \mid \mu, \sigma^2)$ denotes the density of the normal distribution. Assuming the prior distribution $\mu \sim N(\theta_0, \tau^2)$, where θ_0 and τ^2 are hyperparameters, the posterior distribution of μ is derived as

$$
\begin{aligned}
f(\mu \mid D) \quad &\propto \quad f(\mu \mid \theta_0, \tau^2) L(D \mid \mu) \\
&\propto \quad \exp\left\{ -\frac{1}{2\tau^2}(\mu - \theta_0)^2 \right\} \exp\left\{ -\frac{1}{2\sigma^2} \sum_{i=1}^{n} (y_i - \mu)^2 \right\} \\
&\propto \quad \exp\left\{ -\frac{\left(\mu - \frac{\sigma^2 \theta_0 + \tau^2 \sum_{i=1}^{n} y_i}{\sigma^2 + n\tau^2} \right)^2}{\frac{2\tau^2 \sigma^2}{\sigma^2 + n\tau^2}} \right\} \\
&= \quad N\left(\mu \mid \frac{\theta_0/\tau^2 + \sum_{i=1}^{n} y_i/\sigma^2}{1/\tau^2 + n/\sigma^2}, \frac{1}{1/\tau^2 + n/\sigma^2} \right).
\end{aligned}
$$

As can be seen, the posterior mean is the weighted average of the prior mean and the observed sample mean, weighted by the respective variance.

In Example 1.1 or 1.2, the posterior distribution is in the same distributional family as the prior distribution, greatly facilitating the posterior computation. This type of prior is called a *conjugate prior*. For many statistical models, however, the conjugate prior does not exist. In these cases, the posterior distribution does not have a form of standard distributions, and the inference on θ can be made by drawing posterior samples of θ from $f(\theta \mid D)$ via Markov chain Monte Carlo (MCMC) methods, such as the Gibbs sampler or Metropolis-Hastings algorithms. More details about MCMC methods are covered in Bayesian textbooks, including Gelman et al. (2013), Carlin and Louis (2008), and Robert and Casella (2013).

1.1.2 Bayesian inference

According to Bayes' theorem, the posterior distribution contains all relevant information we know about the parameter of interest θ in light of the data and prior knowledge. Once the posterior distribution is available, Bayesian inference about θ can be readily made by exploiting this distribution.

Point and interval estimations

For illustrative purposes, we assume that θ is univariate. Based on the posterior distribution $f(\theta \mid D)$, commonly used Bayesian point estimators include:

- Posterior mean: $\hat{\theta} = \mathrm{E}(\theta \mid D) = \int \theta f(\theta \mid D)\mathrm{d}\theta$;

- Posterior median: $\hat{\theta}$ such that $\int_{-\infty}^{\hat{\theta}} f(\theta \mid D)\mathrm{d}\theta = 0.5$;

- Posterior mode: $\hat{\theta} = \arg\max_{\theta} f(\theta \mid D)$.

When a non-informative prior or the flat prior $f(\theta) \propto 1$ is used, the posterior mode estimator is often equal or similar to the frequentist maximum likelihood estimator.

Because Bayesian statistics treat the unknown parameter θ as a random variable, it is straightforward to make interval inference about θ based on its posterior distribution. Analogous to the frequentist confidence interval, the $100(1-\alpha)\%$ Bayesian credible interval $\mathcal{C}_\alpha$ is given by

$$\int_{\mathcal{C}_\alpha} f(\theta \mid D)\mathrm{d}\theta = 1 - \alpha,$$

where α is a prespecified value such as 0.05. Unlike the frequentist confidence interval, whose interpretation relies on repeated sampling (i.e., repeat the trial many times), the Bayesian credible interval has a more intuitive probability interpretation: the $100(1-\alpha)\%$ Bayesian credible interval is the interval that has a $100(1-\alpha)\%$ chance to contain the true value of θ.

There are different ways to construct credible intervals. A commonly-used one is the equal-tailed credible interval $\mathcal{C}_\alpha = (\theta_{\alpha/2}, \theta_{1-\alpha/2})$, where $\theta_{\alpha/2}$ and $\theta_{1-\alpha/2}$ satisfy

$$\int_{-\infty}^{\theta_{\alpha/2}} f(\theta \mid D)\mathrm{d}\theta = \alpha/2, \quad \text{and} \quad \int_{-\infty}^{\theta_{1-\alpha/2}} f(\theta \mid D)\mathrm{d}\theta = 1 - \alpha/2.$$

Alternatively, the highest posterior density (HPD) interval gives the shortest length among all credible intervals with a given α. A $100(1 - \alpha)\%$ HPD interval for θ is given by

$$\mathcal{C}_\alpha = \{\theta : f(\theta \mid D) \geq \pi_\alpha\},$$

where π_α is chosen such that

$$\Pr(\theta \in \mathcal{C}_\alpha \mid D) = \int_{\theta : f(\theta|D) \geq \pi_\alpha} f(\theta \mid D)\mathrm{d}\theta = 1 - \alpha.$$

When the posterior function $f(\theta \mid D)$ is unimodal and symmetric, the equal-tailed credible interval coincides with the HPD interval. If the posterior function has more than one mode, then it is possible that the HPD interval is a union of several disjointed regions.

In addition, under the Bayesian paradigm, it is straightforward to estimate interval probabilities that are of practical interest and importance, e.g., $\Pr(\theta > \phi)$ where ϕ is a pre-specified cutoff. For example, in Example 1.1, investigators may be interested in estimating the probability that the response rate θ is greater than 0.4, or less than 0.2, which can be useful for treatment decisions and easily computed under the Bayesian framework. However, these quantities cannot be computed under the frequentist framework because the unknown parameter is assumed to be fixed and does not have a distribution.

Prediction

Another strength of Bayesian statistics is its ability to make predictions on future observations. Assuming that the future observations $\tilde{D}$ and the current data D are conditionally independent given θ, the posterior predictive distribution for $\tilde{D}$ is given by

$$
\begin{aligned}
f(\tilde{D} \mid D) &= \int f(\tilde{D}, \theta \mid D)\mathrm{d}\theta \\
&= \int f(\tilde{D} \mid \theta, D)f(\theta \mid D)\mathrm{d}\theta \\
&= \int L(\tilde{D} \mid \theta)f(\theta \mid D)\mathrm{d}\theta.
\end{aligned}
$$

Here, $f(\tilde{D}, \theta \mid D)$ denotes the joint probability function of $\tilde{D}$ and θ given D, and $f(\tilde{D} \mid \theta, D)$ denotes the density function of $\tilde{D}$ given θ and D. The last equation shows that the posterior predictive distribution of $\tilde{D}$ is formed by averaging the distribution of $\tilde{D}$, given a value of θ, over all possible values of θ under its posterior distribution $f(\theta \mid D)$.

Bayesian hypothesis testing

The cornerstone of Bayesian hypothesis testing is the *Bayes factor*. Consider two candidate models (or hypotheses) H_0 and H_1, and let $L(D \mid \theta_k, H_k)$

denote the likelihood function, characterized by model parameters θ_k, under model H_k, $k = 0, 1$. In some applications, it is possible that the competing models have the same set of parameters but this is not necessary. Given a prior distribution $f(\theta_k \mid H_k)$, the marginal likelihood given model H_k, which is the probability of the observed data D given model H_k, is obtained by integrating out θ_k from the joint distribution of (D, θ_k) given H_k, that is,

$$\Pr(D \mid H_k) = \int f(D, \theta_k \mid H_k)\mathrm{d}\theta_k = \int L(D \mid \theta_k, H_k)f(\theta_k \mid H_k)\mathrm{d}\theta_k,$$

for $k = 0, 1$. The Bayes factor is defined as the ratio of the marginal likelihoods under H_1 verus H_0

$$\mathrm{BF}_{10} = \frac{\Pr(D \mid H_1)}{\Pr(D \mid H_0)}.$$

Let $\Pr(H_k)$ denote the prior probability of H_k being true, the posterior probability of H_k being true is given by

$$\begin{aligned} \Pr(H_k \mid D) &= \frac{\Pr(D \mid H_k)\Pr(H_k)}{\Pr(D)} \\ &= \frac{\Pr(D \mid H_k)\Pr(H_k)}{\Pr(D \mid H_0)\Pr(H_0) + \Pr(D \mid H_1)\Pr(H_1)}. \end{aligned}$$

Then, the posterior odds in favor of H_1 versus H_0, given D, is given by

$$\frac{\Pr(H_1 \mid D)}{\Pr(H_0 \mid D)} = \frac{\Pr(H_1)}{\Pr(H_0)}\mathrm{BF}_{10}.$$

Therefore, the Bayes factor can be interpreted as a multiplication factor that converts the prior odds of candidate models being true (i.e., $\Pr(H_1)/\Pr(H_0)$) into the posterior odds of candidate models being true. When the two models are equally probable *a priori* with $\Pr(H_0) = \Pr(H_1) = 1/2$, the Bayes factor reduces to the posterior odds. In other words, the Bayes factor measures the evidence provided by the observed data in favor of one model against the other. A value of $\mathrm{BF}_{10} > 1$ indicates that H_1 is more likely to be true than H_0, given the observed data D, compared to the belief before seeing the observation. In the context of hypothesis testing, Kass and Raftery (1995) provided interpretations of the Bayes factor on the $\log_{10}$ scale as a measure to quantify the level of evidence, see Table 1.1.

1.2 Bayesian adaptive designs

Traditionally, clinical trials are often conducted in a "static" way by fixing design parameters in advance (e.g., the sample size, population to be enrolled,

TABLE 1.1: The interpretation of the Bayes factor as a measure to quantify the level of evidence according to Kass and Raftery (1995).

$\log_{10} \mathrm{BF}_{10}$	BF_{10}	Evidence against H_0
0 to 1/2	1 to 3.2	Not worth more than a bare mention
1/2 to 1	3.2 to 10	Substantial
1 to 2	10 to 100	Strong
> 2	> 100	Decisive

drugs to be tested, treatment assignment, and randomization probability). Such static trials are designed based on a set of (strong) assumptions on the characteristics of the drug and target population (e.g., safety, treatment effect size, and sensitive patients), which unfortunately often turn out to be incorrect. This leads to a high failure rate, loss in efficiency, and high development costs (Berry, 2003).

Adaptive designs provide an important approach to addressing this issue. Per the US Food and Drug Administration (FDA) guidance on Adaptive Designs for Clinical Trials of Drugs and Biologics (FDA, 2019), an adaptive design is defined as a clinical trial design that allows for prospectively planned modifications on one or more aspects of the design based on accumulating data from subjects in the trial. Examples of the modifications include adaptive dose escalation and de-escalation decisions, futility/efficacy/safety monitoring to early terminate or graduate a treatment arm, adding new treatment arms, adaptive randomization based on patients' treatment outcomes or covariates, enriching or reducing the enrollment of a subpopulation during the trial, adaptive treatment allocation, sample size recalculation, seamless phase transition, making trial go/no-go decisions in multistage trials, and adaptive estimation of the treatment effect or toxicity rate using interim data (Zang and Lee, 2014; Pallmann et al., 2018). These in-trial adaptations make the design more accurately reflect the characteristics of the drug effect on the target population, thus potentially improving the safety, efficiency, and success rate of clinical trials (Chow and Chang, 2008; Mahajan and Gupta, 2010; Pallmann et al., 2018).

While adaptive designs can be developed using either frequentist or Bayesian approaches, the Bayesian paradigm is particularly appealing due to its "learn-as-we-go" nature. As described previously, by repeatedly applying Bayes' theorem, the Bayesian paradigm seamlessly integrates accumulating interim trial data and updates the knowledge of the parameter of interest θ to guide treatment decisions. Compared to frequentist approaches, Bayesian methods have several advantages (Lee and Chu, 2012; Berry et al., 2010), including:

1. Bayesian designs treat the parameters of interest (e.g., the treatment effect) as random variables, directly modeled by well-defined probability distributions. Trial decisions and inference are made based on the (posterior or predictive) probability distribution of the parameters of interest, thus they properly incorporate various levels of uncertainty contained in the data.

2. Because of the probability-based inferential structure, Bayesian methods directly answer the scientific questions through the use of posterior distributions, e.g., the probability that the response rate of the drug is greater than a certain benchmark. This facilitates communication with physicians and provides a consistent way to tackle more complex problems.

3. Bayesian methods support continuous learning in a learn-as-we-go fashion, which forms the basis of adaptive clinical trial designs. As long as new data or any relevant information is available during the trial, the design can be automatically integrated in the posterior inference at predefined time points for making interim decisions. This feature greatly facilitates flexible and efficient trial adaptation and monitoring in adaptive clinical trials, without compromising the integrity and validity of the trial.

4. As an essential feature of Bayesian approaches, prior or external information can be naturally synthesized into the current trial. With a properly specified prior distribution, Bayesian adaptive designs tend to make more accurate and efficient decisions, which has the potential to result in a trial with less cost, a reduced sample size, and a shorter duration.

5. By applying outcome adaptive randomization, more patients can be treated by the putatively more effective treatment(s) based on the interim data. For life-threatening diseases, this is a desirable feature. Before a definitive decision is made, the probability of patient allocation to arms can be computed to maximize the overall benefit, e.g., the overall response rate, in the trial and/or the future patient horizon in the population while controlling the false positive and false negative error rates in decision making.

6. The Bayesian framework provides an ideal tool for hierarchical modeling. Different levels of information can be naturally incorporated under the hierarchical model assumption, which enables information borrowing across different patient subgroups, treatment regimes, and time periods. Such a feature is of great importance in designing basket, umbrella, or platform trials in the modern era of precision medicine.

7. The "gain/loss" utility functions can be formally incorporated by Bayesian approaches, which facilitates informed decision making in

a complex setting of balancing risk and benefit. Optimal treatment or trial decisions can be obtained by maximizing the utility function (or minimizing the loss function) in the Bayesian decision-theoretic framework. Furthermore, the posterior predictive distribution renders the evaluation of future patient outcomes based on the current observed data, which offers additional flexibility in adaptively treating future patients to maximize risk–benefit tradeoffs.

1.3 Adoption of Bayesian adaptive designs

Bayesian adaptive designs are gaining popularity in clinical research, especially in oncology. As one of the world's largest premier cancer centers, The University of Texas MD Anderson Cancer Center (MDACC) has pioneered the application of Bayesian adaptive designs in clinical trials. Since the early 2000s, Bayesian approaches have become an indispensable approach in designing clinical trials at MDACC. Biswas et al. (2009) and Tidwell et al. (2019) reviewed a total of 1984 trials (964 and 1020 trials, respectively) registered in the MDACC electronic protocol database from January 2000 to April 2005, and from January 2009 to December 2013. Table 1.2 presents the numbers and percentages of Bayesian and non-Bayesian trials by site, respectively. Bayesian trials were defined as trials which incorporated at least one Bayesian feature in the design. Of the 964 trials registered during 2000 to early 2005, 394 (41%) were multicenter trials and 570 (59%) were MDACC-only trials. Among the MDACC-only trials, 30% (169 out of 570) implemented Bayesian designs between 2000 and 2005; the percentage of Bayesian trials significantly increased to 56% (189 out of 335) between 2009 and 2013. Such an increasing trend in use of Bayesian designs was also observed in multicenter trials: the percentage of Bayesian trials conducted in 2009–2013 (14%) doubled that in 2000–2005.

Figure 1.2 depicts the number of protocols by year and development phase for the two time periods. Panels (a) and (b) of Figure 1.2 demonstrate the increasing use of Bayesian designs over time: the percentage of trials that applied Bayesian methods grew from 16% in 2000 to 30% in 2013. Panels (c) and (d) of Figure 1.2 show that the Bayesian designs were more commonly adopted in early-phase trials, such as phase I, phase II, and seamless phase I–II trials. This is expected as early-phase trials are explorative in nature, starting with substantial unknowns. Bayesian adaptive designs allow investigators to efficiently synthesize accumulating data and make timely adaptions and modifications to treat patients and learn about the treatment more efficiently (Lin and Lee, 2020). Furthermore, due to less stringent regulatory requirements for early-phase trials as compared to late-phase trials, e.g., no need a rigorous control of type I/II errors, implementation of Bayesian designs in early-phase clinical trials is more acceptable. This partially explains the lower percentage

TABLE 1.2: Numbers and percentages of Bayesian and non-Bayesian trials by site.

Site	Bayesian	Non-Bayesian	Total
Period: Janauary 2000 to April 2005			
MDACC only	169 (30%)	401 (70%)	570
Multicenter	26 (7%)	368 (93%)	394
All sites	195 (20%)	769 (80%)	964
Period: January 2009 to December 2013			
MDACC only	189 (56%)	146 (44%)	335
Other sites only	0 (0%)	9 (100%)	9
Multicenter	94 (14%)	582 (86%)	676
All sites	283 (28%)	737 (72%)	1020

of multicenter trials using Bayesian designs, as shown in Table 1.2, because multicenter trials are usually larger, late-phase trials.

Despite the encouraging trend of using Bayesian adaptive designs, some major barriers remain. Compared to conventional designs, Bayesian adaptive designs are often difficult to understand, require complicated statistical modeling, demand complex computation and simulations to explore the operating characteristics, and need expensive infrastructure for implementation. As a result, in most research institutions and pharmaceutical industry, the use of Bayesian adaptive designs is still limited. To explore the promising features of the Bayesian adaptive designs, FDA launched the Complex Innovative Trial Design (CID) Pilot Program in 2018. Through this initiative, FDA encouraged the collaboration among industry, academia, and regulatory agency to learn and advance novel clinical trial methodology.

In the last decade, great progress has been made to break these barriers. One direction that holds great promise and potential is the development of model-assisted designs (Yuan et al., 2019). Model-assisted designs yield superior performance compared to conventional algorithm-based designs, and their decision rules can be pre-tabulated and included in the protocol and thus implemented as simply as conventional designs. In the remaining chapters of this book, we will introduce and review model-assisted designs to address the quandary of simplicity versus performance that hinders the adoption of Bayesian adaptive designs in practice. Software and trial examples will be provided to illustrate the methodology. We consider that novel model-assisted

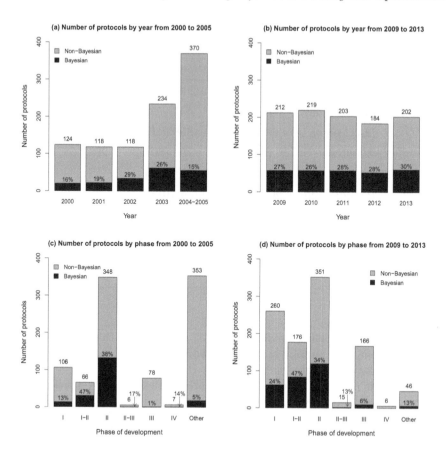

FIGURE 1.2: Number of trial protocols and percentage of Bayesian trials (highlighted by shaded black areas) by year and phase of development. The number above the bar represents the total number of trial protocols, and the percentage shows the percentage of Bayesian trials.

designs possess features where superiority meets simplicity. They adhere to the new KISS principle: Keep it simple and smart! The overarching goal of this book is to increase the use of Bayesian adaptive designs, thereby speeding up and improving the success of early phase clinical trials.

2

Algorithm-Based and Model-Based Dose Finding Designs

2.1 Introduction

Conventionally, the objective of a phase I clinical trial is to identify the highest dose of a new drug that is acceptably safe. This dose is called the maximum tolerated dose (MTD), typically defined as the dose with the probability of causing a dose-limiting toxicity (DLT) closest to a prespecified target rate, for example 20% or 30%. The adverse events (AEs) that define a DLT are prespecified by the investigators, and often scored using the Common Terminology Criteria for Adverse Events (CTCAE) from the National Cancer Institute (NCI). The CTCAE defines the severity of the AE using a 5-grade scale based on the general guideline: grade 1 is mild; grade 2 is moderate; grade 3 is severe or medically significant but not immediately life-threatening; grade 4 is life-threatening; and grade 5 is death related to AE. Typically, the DLT is often defined as an AE of grade 3 or higher.

An implicit assumption for finding the MTD is that efficacy and toxicity of the investigational drug both increase with the dose, and thus the MTD is presumed to be the most efficacious dose with an acceptable probability of causing a DLT. This dose–toxicity–efficacy monotonicity assumption is reasonable for most conventional cytotoxic agents, such as chemotherapies. For many novel molecularly targeted or immunotherapy agents, efficacy may not monotonically increase with the dose, although toxicity generally increases with the dose. In these cases, a more appropriate objective for dose-finding trials is to find the optimal biological dose (OBD), which is generally defined as the dose that has the highest desirability in terms of the efficacy-toxicity tradeoff. For example, assume that 20 mg of a new drug produces the efficacy probability of 0.4 and toxicity probability of 0.3, if 10 mg produces the efficacy probability of 0.39 and toxicity probability of 0.1, then 10 mg is more desirable because it yields comparable (or higher) efficacy with lower toxicity. Chapter 8 describes designs for finding the OBD.

Due to logistical reasons (e.g., preparation and manufacture of the drug), the set of doses to be explored in a phase I trial is often prespecified by investigators. The lowest dose typically is specified as one-tenth of the dose that killed 10% of rodents during pre-clinical studies, i.e, one-tenth of the LD_{10}

DOI: 10.1201/9780429052781-2

13

in rodents, after adjusting for differences in body surface area. The other doses may be specified to follow a modified Fibonacci sequence so that successive increments are, say 100%, 67%, 50%, 40%, and 33% thereafter. Alternatively, successive increments may be a fixed percentage, e.g., 33%, or quantity, e.g., 50 mg.

Two central statistical issues in dose finding are

1. Dose exploration (i.e., dose escalation/de-escalation), that is, how to explore the set of doses during the trial?

2. Dose selection, i.e., how to determine the MTD upon completion of the trial?

One unique challenge associated with these issues is that we should consider not only statistical efficiency, but also patient ethics. On one hand, dose exploration should proceed quickly through doses that are well below the MTD, since these doses presumably are subtherapeutic. On the other hand, dose exploration in the phase I trial should also proceed cautiously to avoid exposing an excessive number of patients to doses that are above the MTD.

Dose selection is critical because the dose selected as the MTD upon completion of the phase I trial will likely be used in subsequent phase II and III trials. If the dose selected as the MTD is well below the true MTD, then a subsequent trial may likely fail, thereby wasting enormous resources and possibly overlooking a promising new drug. If the dose selected as the MTD is well above the true MTD, then a subsequent trial will expose patients to an unsafe dose and likely be terminated early due to an excessive rate of DLTs.

Phase I trial designs can be generally classified as algorithm-based designs, model-based designs, and model-assisted designs (Yuan et al., 2019). This taxonomy is based more on the characteristics of implementation of designs, rather than statistical theory, to facilitate practitioners to understand and apply the designs.

Algorithm-based designs use a set of simple, pre-specified rules (or algorithms) to determine dose escalation and de-escalation, without assuming any model on the dose–toxicity relationship. Examples include the 3+3 design (Storer, 1989), the accelerated titration design (Simon et al., 1997), the rolling-six design (Skolnik et al., 2008), the A+B design (Lin and Shih, 2001), the biased-coin design (Durham et al., 1997) and their variations (Ivanova et al., 2003; Stylianou and Follmann, 2004). The implementation of the designs does not require a computer program or much support from statisticians. Despite widespread criticism of the 3+3 design for poor operating characteristics, its simplicity continues to make it one of the most widely used phase I trial designs in practice.

In contrast to the algorithm-based designs, model-based designs utilize prespecified parametric dose–toxicity models (e.g., the power model or logistic model) to guide dose escalation and de-escalation. As information accrues during the trial, the dose–toxicity relationship is re-evaluated by updating the estimates of the model parameters and then used to guide the dose allocation for subsequent patients. A typical example of the model-based design is the continuous reassessment method (CRM) (O'Quigley et al., 1990). Although a model-based design, such as the CRM, yields better performance than an algorithm-based design (Le Tourneau et al., 2009; Jaki et al., 2013; van Brummelen et al., 2016), it is considered by many to be statistically and computationally complex due to the requirement of specifying the model and prior, as well as repeated model fitting and estimation. This leads practitioners to perceive dose allocations as coming from a "black box." As a result, the use of the model-based designs has been fairly limited in practice (Rogatko et al., 2007).

Emerging in the last decade, model-assisted designs combine the simplicity of algorithm-based designs and the good performance of model-based designs. This class of designs utilizes a probability model to derive the design, similar to the model-based designs, but their rules of dose escalation and de-escalation can be pre-tabulated before the onset of the trial in a fashion similar to the algorithm-based designs. Examples of model-assisted designs include the modified toxicity probability interval (mTPI) design (Ji et al., 2010), Bayesian optimal interval (BOIN) design (Liu and Yuan, 2015; Yuan et al., 2016a), and keyboard design (Yan et al., 2017). Due to their competitive performance and simplicity, model-assisted designs have been increasingly used in practice (Yuan et al., 2019).

Statistically, these three classes of designs are more or less intertwined. For example, algorithm-based designs involve, explicitly or implicitly, a certain probability model (e.g., the binomial or Bernoulli model). Model-based designs often also use some algorithms/rules to guide dose escalation (e.g., no skipping of untried doses). Model-assisted designs, from a certain perspective, might be regarded as a hybrid of model-based and algorithm-based designs as their decision rules are model-based but can be enumerated and implemented in a way similar to algorithm-based designs.

In this chapter, we briefly overview some algorithm-based designs and model-based designs to lay down the foundation for the model-assisted designs, the focus of this book and subsequent chapters. We assume that the DLT is scored as a binary variable (i.e., DLT/no DLT) and is quickly ascertainable such that when a new patient is enrolled and ready for dose assignment, the DLT outcomes have been ascertained for all patients already enrolled. In Chapters 5 and 7, we will discuss how to design phase I trials when the DLT cannot be ascertained quickly (i.e., late-onset) and account for toxicity grades scored in a scale of more than two levels.

Before describing the designs, we establish some notation. Let $d_1 < \cdots < d_J$ denote the J prespecified doses of the new drug that is under investigation in the trial. We use $\pi(d_j)$, or shorthand π_j when no confusion is caused, to denote the DLT probability that corresponds to d_j, and ϕ to denote the target DLT probability for the MTD. We use n_j to denote the number of patients who have been assigned to d_j, and y_j to denote the number of DLTs observed at $d_j, j = 1, \ldots, J$. Therefore, at a particular point during the trial, the observed data are $D = \{D_j, j = 1, \ldots, J\}$, where $D_j = (n_j, y_j)$ are the "local" data observed at dose level j.

2.2   Traditional 3+3 design

The traditional 3+3 design is the most widely used algorithm-based phase I trial design in practice. This design is transparent and easy to implement, but has poor operating characteristics (O'Quigley and Chevret, 1991; Le Tourneau et al., 2009) and suffers from a number of drawbacks. For example, it cannot target a specific toxicity rate, has poor accuracy in identifying the MTD, and tends to treat a large percentage of patients at low doses that are potentially subtherapeutic.

The so-called 3+3 design actually is a family of designs, including numerous variations. Figure 2.1 shows the dose escalation and de-escalation rules for a commonly used version of the 3+3 design. The design sequentially treats patients in cohorts of three, and typically defines the MTD as the highest dose, at which no more than one out six patients has the DLT. The toxicity outcomes in the current cohort must be observed before any patients in the next cohort enter the trial.

Under the 3+3 design depicted in Figure 2.1(a), the first cohort is treated at a prespecified starting dose, which is usually the lowest dose under consideration in the trial, and each subsequent cohort is treated as follows:

- If 0/3 patient in the current cohort has a DLT, the next cohort is treated at the next higher dose.

- If 1/3 patient has a DLT, the next cohort is treated at the same dose.

- If $\geq 2/3$ patients have a DLT, the current dose and all higher doses are declared to be above the MTD, and the next cohort is treated at the next lower dose. In the case that the current dose is the lowest dose, then the trial is terminated and no dose is selected as the MTD.

After two cohorts have been treated at the same dose, i.e., six total patients, the design proceeds as follows:

- If $\leq 1/6$ patients had a DLT, the next cohort is treated at the next higher dose, provided that this dose has not previously been declared to be above the MTD.

- If $\geq 2/6$ patients present with a DLT, then the next lower dose is selected as the MTD, provided that 6 patients have already been treated in that dose.

There are different versions of the 3+3 design. A more aggressive variation of the traditional 3+3 design is to select the MTD as the highest dose at which two or fewer of six patients present with a DLT, see Figure 2.1(b).

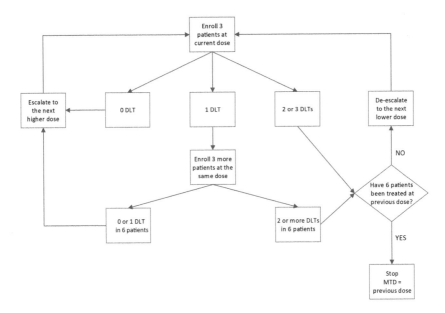

(a) A 3+3 design targeting the MTD with the DLT rate $\leq 1/6$

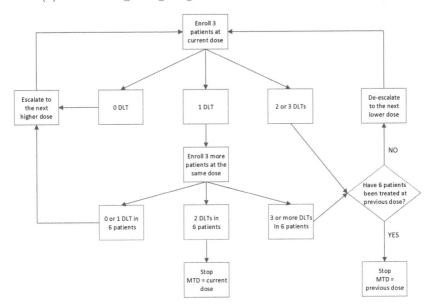

(b) A 3+3 design targeting the MTD with the DLT rate $\leq 2/6$

FIGURE 2.1: Two different versions of the 3+3 design.

Many numerical studies show that the 3+3 design has poor operating characteristics, e.g., poor accuracy to identify the MTD (Ahn, 1998; Iasonos et al., 2008; Onar-Thomas and Xiong, 2010; Zhou et al., 2018b). This can be explained by Table 2.1 applying the design shown in Figure 2.1(a), which displays each of the possible outcomes and the corresponding posterior summaries, including the 95% credible interval of the DLT probability and the posterior probability that the DLT probability is greater than 25%. The posterior summaries are derived by assuming independent beta-binomial models for π_j at each d_j, i.e.,

$$y_j \mid n_j, \pi_j \quad \sim \quad \text{Binomial}(n_j, \pi_j)$$
$$\pi_j \quad \sim \quad \text{Beta}(0.5, 0.5),$$

resulting in the posterior

$$\pi_j \mid y_j, n_j \overset{iid}{\sim} \text{Beta}(y_j + 0.5, n_j - y_j + 0.5).$$

The 3+3 design does not target a specific DLT probability, but a range of DLT probabilities approximately ranging from 1/6 to 1/3, with the mean of 0.25. Thus, $\text{Prob}(\pi_j \leq 0.25 \mid y_j, n_j)$ is used to summarize the posterior evidence for whether d_j is tolerable.

As shown in Table 2.1, the widths of the 95% credible intervals, which are mostly > 0.5, indicate that three or six patients are far from adequate for precisely estimating the DLT probability at a particular dose. For example, for a dose with 1/6 DLT, the 95% credible interval for the true DLT probability is (0.019, 0.558), which is too wide to make meaningful assessment of safety. In addition, despite that doses corresponding to the outcome 2/6 still have

TABLE 2.1: What can be learned from the traditional 3+3 design?

No. DLTs/No. Patients	$\text{Prob}(\pi \leq 0.25 \mid \text{Data})^\dagger$	95% Credible Interval†
Dose is considered to be below the MTD		
0/3	0.830	(0.000, 0.536)
1/6	0.652	(0.019, 0.558)
Dose is selected as the MTD		
0/6	0.942	(0.000, 0.330)
1/6*	0.652	(0.019, 0.558)
Dose is declared to be above the MTD		
2/6	0.298	(0.077, 0.714)
3/6	0.085	(0.167, 0.833)
2/3	0.058	(0.177, 0.961)
4/6	0.014	(0.286, 0.923)
3/3	0.003	(0.464, 1.000)

*When the next higher dose has been used to treat patients. †Assuming that the prior distribution of the DLT probability is Beta(0.5, 0.5)

approximately 30% chance to be at or below the MTD (i.e., $\pi_j \leq 25\%$), the traditional 3+3 design will declare any dose with this outcome to be above the MTD.

These results highlight that the fundamental problem of the 3+3 design is not its dose escalation and de-escalation rules, which actually are reasonable when the target DLT probability is around 25%, but that it often stops too early at a dose lower than the MTD and unreliably claims the MTD based on merely six patients treated at that dose. In addition, its algorithm precludes the possibility of calibrating the sample size to achieve higher reliability and better operating characteristics. As these issues are not related to the number of doses, the common belief that the 3+3 design works well for a trial with a few (e.g., two or three) doses actually is incorrect. Due to the same reason, the extension of the 3+3 design, such as the rolling six design (Skolnik et al., 2008), also has poor accuracy in identifying the MTD (Onar-Thomas and Xiong, 2010; Zhao et al., 2011). Furthermore, algorithm-based designs are rigid and inflexible when the study conduct deviates from the design. For example, a decision cannot be rendered for the 3+3 design if two or four patients are treated at a certain dose. This rigidity poses tangible operational problems as clinical trials often might not be carried out exactly as designed.

2.3 Cohort expansion

A common approach to address the sparse information at the MTD, selected by the 3+3 design, is to perform cohort expansion at the MTD. That is, treating an additional certain number of patients (e.g., 6 or 10) at the estimated MTD to confirm safety. This approach may seem sensible, since a larger sample size provides a more reliable inference, but actually suffers from a number of logical, scientific, and ethical flaws.

The practice of adding an expansion cohort after selecting the MTD is based on the fallacious assumption that the MTD selected by the 3+3 design is reliable to be the true MTD. This ignores the basic statistical principle that any estimate computed from a small sample size is associated with large uncertainty. Suppose that a dose is selected as the MTD, at which 1/6 patients had DLT, the 95% credible interval (see Table 2.1) shows there is a 95% chance that the true toxicity probability of that dose is between 0.019 and 0.558. As a result, a wide variety of questions may arise for the different numbers of toxicities that may be observed in the expansion cohort. Since six is a very small sub-sample, as more patients are treated at the estimated MTD in an expansion cohort, the additional toxicity data easily may contradict the earlier conclusion that the selected dose is the MTD. For example, what should one do if the first three patients in an expansion cohort of size 10 all have toxicity?

This would give a total of $1/6 + 3/3 = 4/9$ (44%) toxicities at the MTD. The initial 3/3 toxicities in the expansion cohort suggest that the selected MTD is not safe. Should one treat seven more patients at the estimated MTD, as mandated by the protocol, or violate the protocol by abandoning the MTD and de-escalating to a lower dose? If one de-escalates, what sort of rule or algorithm should be applied to choose a dose, or doses, for the remaining seven patients? If one continues to treat patients at the selected MTD, and ends up with 7/10 toxicities, for a total of $1/6 + 7/10 = 8/16$ (50%), what should one conclude? The point is that the idea of treating a fixed expansion cohort at a chosen MTD may seem sensible, but in practice can be problematic.

In recent years, the sizes of expansion cohorts following phase I trials have exploded, from 6 or 10 to hundreds in some protocols (Bugano et al., 2017). What nominally is a large "phase I expansion cohort" actually is a phase II trial, but conducted without any design, other than a specification of sample size. This practice magnifies all of the problems described above that occur with a small expansion cohort. It fails to use the new data in the expansion cohort adaptively to change the MTD if appropriate, and thus fails to protect patient safety adequately. Furthermore, expansion cohorts are often used to provide an initial exploration of treatment efficacy in specific subgroups. However, without a statistical design, there is no provision to discontinue trials when experimental treatments are ineffective. As a general rule of thumb, if the size of an expansion cohort reaches to the size of a phase II study, e.g., 30, a statistical design is required. More discussion of scientific and ethical pitfalls of cohort expansion can be found in Yan, Thall and Yuan (2017).

2.4   Accelerated titration design

Phase I trials often include many doses that are well below the MTD. This is because the starting dose usually is chosen conservatively based on preclinical studies. As a result, a large number of patients might be treated at subtherepeutic doses that are far below the MTD. To address this issue, Simon et al. (1997) proposed the accelerated titration design (ATD), which includes an initial phase of accelerated dose exploration to quickly transition from these presumably inefficacious doses to the more promising doses. During the initial phase of accelerated titration, patients are treated in the cohort size of one until the first DLT is observed or two moderate (grade 2) toxicities occur. At that point in the trial, two additional patients are treated at that dose, and then dose exploration proceeds as in the traditional 3+3 design. Compared to the traditional 3+3 design, on average adding the accelerated titration leads to fewer patients at the doses below the MTD, but slightly more patients at the doses above the MTD.

2.5 Continual reassessment method

Model-based designs are developed to find the MTD in a more efficient and deliberate way. This class of designs assumes a parametric model between the dose and DLT probability to facilitate the dose exploration in a statistically justified manner. In contrast, the traditional 3+3 design and the ATD rely on simple ad hoc rules for dose exploration that lack clear statistical justification.

The continual reassessment method (CRM; O'Quigley et al. (1990)) is the first example of model-based dose finding designs. The CRM assumes a parametric model to describe the dose–toxicity relationship. As information accrues during the trial, the dose–toxicity curve is re-evaluated by updating the estimates of the model parameters via Bayesian machinery, and then used to guide the dose allocation for subsequent patients. The commonly used model for the CRM is the following power model,

$$\pi_j = q_j^{\exp\{\alpha\}}, \text{ for } j = 1, \ldots, J, \tag{2.1}$$

where α is the unknown parameter, and $q_1 < \cdots < q_J$ are prior estimates of the DLT probabilities, called the skeleton, at each of the dose levels, respectively. Other models, such as single-parameter logistic and hyperbolic tangent models have also been proposed for the CRM, see, Cheung (2011, Section 3.2.2). Research shows that the choice of the model has little impact on the performance of the design. What is more important is the configuration and calibration of the model, e.g., the skeleton and priors.

Because the sample size of phase I trials is typically small, the model used in the CRM is often simple and parsimonious, containing one or two parameters (Iasonos et al., 2016). For illustration, under a target DLT rate of 0.25, Figure 2.2 depicts the family of dose–toxicity curves that the power model covers with the skeleton $(q_1, \cdots, q_6) = (0.01, 0.04, 0.12, 0.25, 0.40, 0.54)$ for a trial with six doses. This skeleton is computed using the model calibration approach of Lee and Cheung (2009) with the half-width of the indifference intervals being 0.07 and the prior guess of the MTD being the fourth dose.

Let $D = \{(n_j, y_j), j = 1, \ldots, J\}$ denote the accrued data with y_j DLTs in n_j patients at dose level j after n patients have been treated in the trial, $n = \sum_{j=1}^{J} n_j$. The likelihood function under (2.1) is

$$L(D \mid \alpha) = \prod_{j=1}^{J} \left[q_j^{\exp\{\alpha\}} \right]^{y_j} \left[1 - q_j^{\exp\{\alpha\}} \right]^{(n_j - y_j)}.$$

Let $f(\alpha)$ denote the prior distribution for α, which often is taken as $N(0, 2)$. Applying Bayes' theorem, the posterior distribution for α is

$$f(\alpha \mid D) = \frac{L(D \mid \alpha)f(\alpha)}{\int L(D \mid \alpha)f(\alpha)\mathrm{d}\alpha}.$$

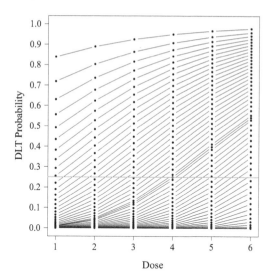

FIGURE 2.2: Prior support of the power model with the skeleton of (0.01, 0.04, 0.12, 0.25, 0.40, 0.54) and a normal $N(0, 2)$ prior distribution assigned to the unknown parameter α.

The posterior mean estimate of the actual DLT probability at dose level j is

$$\widehat{\pi}_j = \int q_j^{\exp\{\alpha\}} f(\alpha \,|\, D) \mathrm{d}\alpha, \text{ for } j = 1, \ldots, J. \tag{2.2}$$

Because α is the only unknown parameter, $\{\widehat{\pi}_j\}$ can be calculated using adaptive quadrature. Alternatively, these can be estimated using a Markov chain Monte Carlo (MCMC) sampling algorithm, but adaptive quadrature is faster, reasonably accurate, and does not require posterior convergence monitoring.

Dose exploration in the CRM is based on the continuously updated estimates of the true DLT probabilities defined in (2.2), and the prespecified target toxicity rate ϕ. The original CRM proposed to treat each cohort at the dose level corresponding to the estimated DLT probability closest to the target ϕ, i.e., $j^* = \arg\min_j\{|\widehat{\pi}_j - \phi|\}$. In practice, for patient safety, a "no-dose-skipping" rule is typically imposed that treats the first cohort of patients at the pre-specified starting dose and prohibits an untried dose to be skipped when escalating the dose (Faries, 1994; Goodman et al., 1995; Piantadosi et al., 1998). Zhou et al. (2018b) show that allowing dose skipping in the CRM substantially increases the risk of overdosing patients with little gain on the accuracy of finding the MTD. Thus, in general we recommend that the no-dose-skipping rule should always be used.

The dose-finding algorithm of the CRM is described below.

1. Patients in the first cohort are treated at the lowest dose d_1 or the pre-specified dose.

2. Based on the cumulated data, we obtain the posterior DLT probability estimates $\hat{\pi}_j$, and find dose level j^* that has a DLT probability closest to the target ϕ. Let j denote the current dose level,

   - if $j^* < j$, de-escalate the dose level to $j - 1$;
   - if $j^* > j$, escalate the dose level to $j + 1$;
   - otherwise, stay at the same level j for the next cohort of patients.

3. The trial continues until the planned maximum sample size N is exhausted, after which the MTD is selected as the dose with $\hat{\pi}_j$ closest to the target ϕ, i.e.,

$$\text{MTD} = \underset{d_j \in (d_1, \ldots, d_J)}{\arg\min} |\hat{\pi}_j - \phi|. \tag{2.3}$$

Although not included in the original CRM, an early stopping rule is often imposed in practice to guard against the situation where the lowest dose is excessively toxic. Specifically, if

$$\Pr(\pi_1 > \phi \mid D) > p_{\text{cut}},$$

then the trial is terminated, where p_{cut} is a prespecified large probability cutoff, e.g., $p_{\text{cut}} = 0.90$. That is, the trial is stopped for safety when the posterior probability that the lowest dose has a DLT probability above the target exceeds a specified threshold, p_{cut}. In addition to the early stopping rule, Heyd and Carlin (1999) proposed to allow the trial to stop early also when the width of the posterior 95% probability interval becomes sufficiently narrow for the dose selected as the MTD. Numerous studies show that the CRM has substantially better performance than the 3+3 design (Ahn, 1998; Garrett-Mayer, 2006; Zhou et al., 2018b).

An important practical issue associated with the CRM is how to specify the skeleton, $q_1 < \cdots < q_J$. If $\alpha = 0$, then $\pi_j = q_j$, for $j = 1, \ldots, J$. Therefore, given the normal prior $f(\alpha) = N(0, 2)$, the skeleton represents the set of prior (median) estimates for the DLT probabilities of J doses, which can be elicited from clinicians. The skeleton determines the initial structure of the model and the family of dose–toxicity curves that the model can cover. As the decisions of dose exploration and selection are made based on the model estimates, different skeletons may lead to quite different design properties, especially under the small sample size of the phase I trials. Yin and Yuan (2009b) showed that the selection probability of the target dose can be 40% lower under one skeleton than that under another skeleton in finite samples. The CRM may perform poorly if it is based on a skeleton such that the true DLT probabilities cannot be approximately recovered by the parsimonious model defined in (2.1). Unfortunately, because the actual DLT probabilities are unknown, practitioners often lack adequate information to determine whether a skeleton is reasonable or not.

To simplify the specification of the skeleton, Lee and Cheung (2009) developed an automatic approach for selecting a skeleton when reliable prior information is lacking. Their method is based on the "indifference interval," which is the range of DLT probabilities that the CRM cannot distinguish from the target ϕ in large samples. The indifference-interval method determines a skeleton based on ϕ (the target toxicity rate), $j^\dagger$ (the prior guess for which dose level is the MTD), J (the number of doses under consideration in the trial), and δ (the desired half-width of the indifference interval). For the power model CRM, the recommended skeleton is

$$
\begin{aligned}
q_j &= \phi, & j &= j^\dagger, \\
q_j &= \exp\left\{ \frac{\log(\phi - \delta)\log(q_{j+1})}{\log(\phi + \delta)} \right\}, & j &= j^\dagger - 1, \ldots, 1, \\
q_j &= \exp\left\{ \frac{\log(\phi + \delta)\log(q_{j-1})}{\log(\phi - \delta)} \right\}, & j &= j^\dagger + 1, \ldots, J.
\end{aligned}
\tag{2.4}
$$

Pan and Yuan (2017) showed that given a fixed half-width of the indifference interval (i.e., δ), the indifference-interval method is invariant to the prior guess for which dose level is the MTD (i.e., $j^\dagger$). That is, different values of $j^\dagger$ result in equivalent skeletons, where equivalent skeletons are defined as skeletons that lead to the same likelihood under the power model (2.1). As a result, the half-width of the indifference interval plays a much more important role than the prior guess of the MTD location.

Lee and Cheung's method simplifies the specification of the skeleton, but does not resolve the sensitivity issue of the CRM pertaining to model misspecification. Table 2.2 shows the simulation results of the CRM with two different skeletons, skeleton 1 = (0.070, 0.127, 0.200, 0.286, 0.377, 0.468) and skeleton 2 = (0.012, 0.069, 0.200, 0.380, 0.560, 0.706), respectively generated by the method of Lee and Cheung (2009) with a half-width of the indifference interval of 0.04 and 0.08. We can see that skeleton 1 substantially outperforms skeleton 2 in scenario 1, whereas the result is opposite in scenario 2. In other words, a skeleton that works well in one scenario may not work well in another scenario, and there does not exist a single "best" skeleton that outperforms all others in every scenario. In the next section, we will describe how to solve this sensitivity issue by specifying multiple skeletons and then using Bayesian model averaging or selection to adaptively identify the best fitted skeleton for robust decision making.

TABLE 2.2: The performance of the CRM and Bayesian model averaging CRM (BMA-CRM) with two different skeletons generated with half-widths of the indifference interval of 0.04 (skeleton 1) and 0.08 (skeleton 2). The target toxicity rate is $\phi = 0.2$ and sample size is $N = 36$.

		1	2	3	4	5	6
				Scenario 1			
Dose level		1	2	3	4	5	6
True DLT rate		0.03	0.04	0.05	0.06	0.07	0.20
CRM (skeleton 1)	% sel[§]	0	0.10	1.25	3.85	21.05	**73.20**
	No. pts[†]	1.5	1.7	2.2	3.5	8.0	18.8
CRM (skeleton 2)	% sel	0.05	1.35	5.70	8.10	28.25	**56.40**
	No. pts	1.5	2.2	3.4	5.2	10.1	13.4
BMA-CRM	% sel	0	0.25	2.45	5.65	21.80	**69.60**
	No. pts	1.5	1.9	2.7	3.9	8.6	17.4
				Scenario 2			
Dose level		1	2	3	4	5	6
True DLT rate		0.12	0.24	0.33	0.60	0.70	0.80
CRM (skeleton 1)	% sel	35.15	**43.50**	10.45	0.25	0	0
	No. pts	13.9	12.2	5.2	1.2	0.3	0.1
CRM (skeleton 2)	% sel	30.60	**53.10**	11.15	0	0	0
	No. pts	12.3	15.1	6.1	1.0	0.1	0
BMA-CRM	% sel	32.30	**49.85**	10.05	0	0	0
	No. pts	12.7	14.3	5.5	1.0	0.3	0.1

[§]: Average selection percentage at each dose;
[†]: Average number of patients treated at each dose.

2.6 Bayesian model averaging CRM

A major practical issue associated with the CRM is the requirement of prespecifying the skeleton. Because of the lack of prior toxicity information on a new drug, physicians may have quite different opinions about which skeleton is reasonable for their particular context. If the true DLT probabilities are not near the prior support of the skeleton, then the CRM may have a high probability of selecting the wrong dose as the MTD. To improve robustness, Yin and Yuan (2009b) proposed prespecifying multiple skeletons, each representing a set of prior estimates of the toxicity probabilities, and using a Bayesian model averaging (BMA) or Bayesian model selection approach to estimate the true DLT probabilities of the doses under consideration in the trial. Yin and Yuan (2009b) showed that the Bayesian model selection approach yields similar performance as the BMA, thus we herein focus on the BMA approach. Through the choice of multiple skeletons, BMA provides a more robust way to construct the dose–oxicity curve compared to the standard CRM.

The rationale behind the BMA-CRM is to use multiple skeletons to represent different dose–toxicity relationships. Each skeleton corresponds to a CRM model described previously with a different set of $q_1 < \cdots < q_J$. As long as one of them is close to the truth, the design will perform well, since BMA automatically identifies and favors the best fitted model.

Specifically, let $\{M_k\}_{k=1}^{K}$ denote the CRM models corresponding to the K prespecified skeletons $\{q_{1k} < \cdots < q_{Jk} : k = 1, \ldots, K\}$. Like the CRM, the kth model in the BMA-CRM connects the actual DLT probabilities to the kth skeleton by assuming,

$$\pi_{jk} = q_{jk}^{\exp\{\alpha_k\}}, \text{ for } j = 1, \ldots, J, \ k = 1, \ldots, K.$$

Let $\Pr(M_k)$ be the prior probability that model M_k is the true model, i.e., the probability that the kth skeleton $(q_{1k} < \cdots < q_{Jk})$ matches the true dose–toxicity curve. If there is no preference *a priori* for any one skeleton over the other skeletons, equal weights can be assigned to the different models by setting $\Pr(M_k) = 1/K$, $k = 1, \ldots, K$. When there is prior information about the importance of each set of the prespecified toxicity probabilities, such information can be incorporated into $\Pr(M_k)$, $k = 1, \ldots, K$. For example, if one skeleton is more likely to be true, it can be assigned a higher prior model probability. After n patients have been treated and the observed data are $D = \{(n_j, y_j), j = 1, \ldots, J\}$, $n = \sum_{j=1}^{J} n_j$, the likelihood function corresponding to the kth model is

$$L(D \mid \alpha_k, M_k) = \prod_{j=1}^{J} \left[q_{jk}^{\exp\{\alpha_k\}} \right]^{y_j} \left[1 - q_{jk}^{\exp\{\alpha_k\}} \right]^{(n_j - y_j)}.$$

The posterior model probability for M_k is

$$\Pr(M_k \mid D) = \frac{m(D \mid M_k)\Pr(M_k)}{\sum_{\ell=1}^{K} m(D \mid M_\ell)\Pr(M_\ell)},$$

where

$$m(D \mid M_k) = \int L(D \mid \alpha_k, M_k) f(\alpha_k) d\alpha_k$$

is the marginal likelihood of model M_k, α_k is the unknown parameter associated with the CRM power model M_k, and $f(\alpha_k)$ is the prior distribution of α_k under model M_k, for $k = 1, \ldots, K$.

During the trial, conditional on the observed data, the models for each skeleton usually yield different estimates of the toxicity probabilities. Some of these estimates may be close to the true values, whereas others may not, depending on how well the models fit the accumulated data. To obtain the estimate of the toxicity probabilities for doses under consideration in the trial, the BMA approach takes a weighted average across the CRM models corresponding to the different skeletons, where the weight for each model reflects how well that model fits the accumulated data relative to the other models. The potential estimation bias caused by a misspecification of the skeleton is averaged out, leading to a more robust design compared to the original CRM. More precisely, the BMA estimate for the toxicity probability at each dose level is given by

$$\bar{\pi}_j = \sum_{k=1}^{K} \hat{\pi}_{jk} \Pr(M_k \mid D), \quad j = 1, \ldots, J, \tag{2.5}$$

where $\hat{\pi}_{jk}$ is the posterior mean of the toxicity probability of dose level j under model M_k, i.e.,

$$\hat{\pi}_{jk} = \int q_{jk}^{\exp\{\alpha_k\}} \frac{L(D \mid \alpha_k) f(\alpha_k)}{\int L(D \mid \alpha_k) f(\alpha_k) \mathrm{d}\alpha_k} \mathrm{d}\alpha_k.$$

By assigning $\hat{\pi}_{jk}$ a weight of $\Pr(M_k \mid D)$, the BMA method automatically identifies and favors the best fitted model, and thus $\bar{\pi}_j$ is always close to the best estimate. Dose allocation and selection in the BMA-CRM is based upon the BMA estimates $\{\bar{\pi}_j\}$, and follows similarly to the original CRM.

For practical use, Yin and Yuan (2009b) recommend using three skeletons in the trial design, and these skeletons should be chosen to reflect different possible shapes of the dose–toxicity curve with different prior locations for the MTD. For example, the first skeleton may increase steadily with the middle dose as the prior MTD, the second skeleton may increase quickly then slow down with a low dose as the prior MTD, and the third skeleton may increase slowly then speed up with a high dose as the prior MTD. The specification of the skeletons should be collaborated with physicians to incorporate their prior knowledge on the possible shapes of the dose–toxicity relationship.

When there is a lack of prior information, in our experience, for $J > 3$, the three skeletons that correspond to $(j^\dagger, \delta) = \{(2, 0.10), (\text{ceiling}((J + 1)/2), 0.07), (J - 1, 0.04)\}$ using the model calibration method of Lee and Cheung (2009) (see Section 2.5) provide a reasonable skeleton set for the BMA-CRM. When there are $J = 3$ doses, the three skeletons can be generated using Lee and Cheung's method with $(j^\dagger, \delta) = \{(1, 0.10), (2, 0.07), (3, 0.04)\}$; and when there are only $J = 2$ doses, the standard CRM with a single skeleton is often sufficient. These skeletons have prior support for a wide range of dose–toxicity curves that are plausible in many contexts. Figure 2.3 depicts the prior support of these three skeletons for a trial with six doses and $N(0, 2)$ prior distributions on α_k, $k = 1, \ldots, K$. In this case, dose levels two, four, and five are the prior expected MTDs for the first, second, and third skeletons, respectively, but it is plausible *a priori* with each of the three skeletons that each dose under consideration is actually the MTD. Alternatively, Pan and Yuan (2017) proposed to choose skeletons by maximizing the model space covered by them,

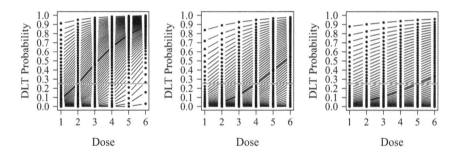

FIGURE 2.3: Prior support of the proposed default set of three skeletons for the BMA-CRM when the target DLT rate is 0.25.

and then further calibrate the skeletons to maximize the probability of correct selection of the MTD using a set of prespecified scenarios. That calibration method is computationally intensive, but it results in a set of three skeletons that performs well in the specified scenarios.

Table 2.2 provides a numerical example that the BMA-CRM is more reliable than the CRM: the performance of the BMA-CRM is close to that of the CRM with the better skeleton in both scenarios 1 and 2. Such robustness and reliability are tremendously important because in practice we often prefer a method that yields reliable performance to a method that has high variability (i.e., performs well in one scenario but not in another scenario).

2.7 Escalation with overdose control

Babb et al. (1998) proposed another model-based design called escalation with overdose control (EWOC). EWOC assumes a two-parameter logistic model for the dose–toxicity curve:

$$\text{logit}\{\pi(d_j)\} = \beta_0 + \beta_1 d_j, \qquad j = 1, \cdots, J, \tag{2.6}$$

where β_0 and β_1 are unknown intercept and slope parameters, respectively, and $\text{logit}(x) = \log\{x/(1-x)\}$. To facilitate interpretation, EWOC re-parameterizes the logistic model by replacing (β_0, β_1) with (π_1, ρ), where $\pi_1 = \pi(d_1)$ is the DLT probability at the lowest dose under consideration in the trial, and ρ is the MTD with $\text{logit}^{-1}(\beta_0 + \beta_1\rho) = \phi$. The re-parameterized logistic model is given by

$$\text{logit}\{\pi(d_j)\} = \frac{(d_j - d_1)\text{logit}(\phi) + (\rho - d_j)\text{logit}(\pi_1)}{\rho - d_1}.$$

If $d_j = \rho$, then $\text{logit}\{\pi(\rho)\} = \text{logit}(\phi)$, and thus ρ is the unknown MTD.

The prior distributions for π_1 and ρ are usually taken to be independent Uniform$(0, \phi)$ and Uniform(d_1, d_J) distributions, respectively. That is, the lowest dose under consideration is assumed to be below the MTD, and the MTD is assumed to be in the investigational dose range $[d_1, d_J]$. An MCMC sampling algorithm can be used for posterior inference. Because of the former model assumption, if a DLT is observed at the lowest dose, then Babb et al. (1998) recommended stopping the trial.

The main difference between the CRM and EWOC is that EWOC explicitly controls the predicted proportion of patients treated with doses that are above the MTD in its dose exploration rule. Specifically, after treating the first cohort at the lowest reasonable dose, EWOC recommends that each subsequent cohort is treated with dose d that has a predicted probability of being above the MTD equal to a prespecified value α,

$$\Pr(\pi(d) \geq \phi \mid D) = \alpha.$$

As $\Pr(\pi(d) \geq \phi \mid D) = \Pr(\rho \leq d \mid D)$, the predicted proportion of patients treated at a dose above the MTD is equal to α. Larger values of α correspond to more aggressive dose escalation. Babb et al. (1998) recommended $\alpha = 0.25$. For patient safety, as with the CRM, the no-dose-skipping rule is often imposed in practice. The trial continues until the prespecified maximum sample size N is exhausted. Upon completion of the trial, EWOC selects the MTD as the dose that minimizes the posterior expected risk

$$\int \mathcal{L}(d; \rho, \alpha) f(\rho|D) \mathrm{d}\rho,$$

with respect to the asymmetric loss function

$$\mathcal{L}(d; \rho, \alpha) = \begin{cases} \alpha(\rho - d), & \text{if } d \leq \rho, \\ (1 - \alpha)(d - \rho), & \text{if } d > \rho, \end{cases}$$

where $f(\rho|D)$ is the marginal posterior function of the MTD. If $\alpha < 0.5$, then $\mathcal{L}(\rho - \delta; \rho, \alpha) < \mathcal{L}(\rho + \delta; \rho, \alpha)$, and thus underdosing is preferable to overdosing. More precisely, since $\mathcal{L}(\rho + \delta; \rho, \alpha)/\mathcal{L}(\rho - \delta; \rho, \alpha) = (1 - \alpha)/\alpha$, this loss function says that the loss incurred by treating a patient with a dose δ units above the MTD is $(1 - \alpha)/\alpha$ times more than treating a patient with a dose δ units below the MTD. Due to the use of the overdose control rule, EWOC is more conservative and safer than the CRM, but at the cost of sacrificing the accuracy of identifying the MTD, which is often substantial. In addition, EWOC is subject to the influence of model misspecification in a similar way as the CRM.

2.8   Bayesian logistic regression method

The Bayesian logistic regression method (BLRM) is another modification of the CRM. The BLRM uses a similar two-parameter logistic regression model as the EWOC, given by

$$\text{logit}\{\pi(d_j)\} = \beta_0 + \beta_1 \log\left(d_j/d^*\right), \qquad j = 1, \cdots, J, \qquad (2.7)$$

where β_0 and β_1 are unknown intercept and slope parameters, d_j is the raw dosage at dose level j, and d^* is the reference dose. Neuenschwander et al. (2008) recommended the use of the vague bivariate normal distribution as the prior distribution of $(\beta_0, \log(\beta_1))$, e.g.,

$$(\beta_0, \log(\beta_1)) \sim N\left(\begin{pmatrix} \mu_1 \\ \mu_2 \end{pmatrix}, \begin{pmatrix} \sigma_1^2 & \rho\sigma_1\sigma_2 \\ \rho\sigma_1\sigma_2 & \sigma_2^2 \end{pmatrix}\right).$$

For example, $(\mu_1, \mu_2, \sigma_1, \sigma_2, \rho) = (-0.847, 0.381, 2.015, 1.207, 0)$.

There are two main differences between the BLRM and the CRM. First, the BLRM requires specifying the proper dosing interval (δ_1, δ_2), defined as the range of the DLT probabilities regarded as acceptable, where $\delta_1 < \phi < \delta_2$. For example, given the target DLT probability $\phi = 0.27$, the proper dosing interval (δ_1, δ_2) may be chosen as (0.20, 0.35). The choice of (δ_1, δ_2) depends on characteristics of the trial (e.g., the target population, toxicity profile of the experimental drug, and risk-benefit consideration), and could differ across trials. Second, the BLRM imposes an overdose control rule, similar to that of the EWOC, as follows: if the observed data suggest that there is 25% posterior probability that the DLT rate of a dose is greater than δ_2, i.e., $\Pr(\pi_j > \delta_2 | D) \geq 0.25$, then that dose is an overdose and cannot be used to treat patients.

The BLRM is highly similar to the EWOC. Both methods use the two-parameter logistic model with slightly different re-parameterization. The main difference is that the BLRM specifies the target as an interval (δ_1, δ_2), while the EWOC uses the point target ϕ as the CRM. However, as the investigational dose is discrete, this difference has little impact on operating characteristics, in particular when setting $\phi = (\delta_1 + \delta_2)/2$. In other words, the two designs will generate similar operating characteristics when the prior is properly calibrated.

To find the MTD, the BLRM proceeds in a similar way as the CRM. The BLRM starts the trial by treating the first cohort of patients at the lowest dose d_1 or a prespecified dose. After each patient cohort is treated, the BLRM updates the estimate of the dose–toxicity curve based on the accumulating DLT data across all dose levels, and assigns the next cohort of patients to the "optimal" dose. The "optimal" dose is defined as the dose j that satisfies the overdose control condition $\Pr(\pi_j > \delta_2 | D) < 0.25$ and meanwhile maximizes the posterior probability of the proper dosing interval (δ_1, δ_2), i.e.,

$\Pr(\pi_j \in (\delta_1, \delta_2)|D)$. For patient safety, as with the CRM, the no-dose-skipping rule is often imposed in practice. Thus, if the estimated optimal dose is higher than the current dose, we escalate the dose by one level; and if the estimated optimal dose is lower than the current dose, we de-escalate the dose by one level. The overdose control rule leads to the following safety stopping rule:

stop the trial if $\Pr(\pi_1 > \delta_2|D) \geq 0.25$ (i.e., the lowest dose is an overdose).

The trial continues until the prespecified maximum sample size N is exhausted. Upon completion of the trial, the BLRM selects the final estimate of the "optimal" dose as the MTD. Alternative early stopping rules can be added to the BLRM, for example, stop the trial if the "optimal" has a large probability (say $\geq 50\%$) of the proper dosing interval or a minimum number of patients have been treated at the "optimal" dose.

Of note, given d_j and the logistic model (2.7), the prior distribution of $\beta_0, \log(\beta_1))$ automatically determines a set of the prior estimates of $\pi(d_j)$, i.e., the skeleton. Thus, the BLRM suffers from the similar sensitivity issue to the skeleton, or more precisely the prior specification. In addition, as noted and elaborated by Cheung (2011) and Iasonos et al. (2016), somewhat counterintuitively, using the more flexible two-parameter logistic model (2.7) actually is inferior to and leads to worse performance than the single-parameter power model (2.1). This is also verified by extensive simulation study by Zhou et al. (2018b), which shows that the CRM (Section 2.5) outperforms the BLRM. Lastly, the BLRM depends on the (standardized) raw dosage d_j, thus may be decision-inconsistent. Consider two drugs with the same number of dose levels, but different raw dosages, e.g., one drug is (5 mg, 10 mg, 15 mg, 20 mg), and the other drug is (30 mg, 60 mg, 120 mg, 240 mg). We call a design decision-consistent if it generates identical operating characteristics as long as the DLT probabilities of the dose levels are the same, regardless of the raw dosages. For example, the two drugs have the same DLT probabilities (0.05, 0.1, 0.25, 0.45) at the four dose levels. Clearly, decision-consistency is a desirable property to have because when the underlying data generation mechanisms are the same, the design should yield the same operating characteristics. The BLRM, however, does not have this property because the same data may result in different estimates when two trials have different raw dosages d_j. In contrast, the CRM and BMA-CRM (based on the power model) do not depend on raw dosages, but only dose levels, and thus are decision-consistent.

2.9 Software

Software is not required to conduct the algorithm-based 3+3 design and the accelerated titration design, since their dose exploration and selection rules are simple and prespecified. In contrast, the model-based designs, such as CRM/BMA-CRM, EWOC, and BLRM, rely on the real-time estimate of dose–toxicity model to make the decision of dose escalation and de-escalation, therefore software is necessary for trial conduct and to obtain simulation-based operating characteristics.

Software for the CRM/BMA-CRM is available in several forms: (1) Windows desktop program "CRM Suite" with intuitive graphical user interface, freely available at MD Anderson Cancer Center Software Download Website `https://biostatistics.mdanderson.org/softwaredownload/`; (2) Web application with intuitive graphical user interface, freely available at `https://www.trialdesign.org`; and (3) R packages freely available at `CRAN`, such as `bcrm` and `dfcrm`.

The EWOC design can be implemented in R using the `bcrm()` function from the `bcrm` package. Interactive software for implementing EWOC is also freely available at `https://biostatistics.csmc.edu/ewoc/index.php`. The BLRM design is available in commercial software such as East by Cytel. R packages such as `blrm` for implementing BLRM can be also found at `CRAN`.

3

Model-Assisted Dose Finding Designs

3.1 Introduction

Model-assisted designs have emerged as an attractive approach for phase I clinical trials that combine the simplicity of algorithm-based designs with the superior performance of model-based designs. Model-assisted designs refer to a class of novel designs that use a model (e.g., the binomial model) for efficient decision making like model-based designs, while their dose escalation and de-escalation rules can be tabulated before the onset of a trial as with algorithm-based designs (Yuan et al., 2019). This chapter introduces several model-assisted phase I designs that aim to find the maximum tolerated dose (MTD), including the modified toxicity probability interval (mTPI) design (Ji et al., 2010), keyboard design (Yan et al., 2017), and Bayesian optimal interval (BOIN) design (Liu and Yuan, 2015; Yuan et al., 2016a).

3.2 Modified toxicity probability interval design

The mTPI design (Ji et al., 2010) starts by defining three dosing intervals: the underdosing interval $(0, \delta_1)$, proper dosing interval (δ_1, δ_2), and overdosing interval $(\delta_2, 1)$, where $\delta_1 < \phi < \delta_2$. For example, given the target probability of the dose-limiting toxicity (DLT) $\phi = 0.2$, the three intervals may be defined as $(0, 0.15)$, $(0.15, 0.25)$, and $(0.25, 1)$, respectively.

Suppose that at the current dose level j, y_j of n_j patients have experienced DLT. mTPI assumes that y_j follows a beta-binomial model

$$
\begin{aligned}
y_j \mid n_j, \pi_j &\sim \text{Binomial}(n_j, \pi_j) \\
\pi_j &\sim \text{Beta}(1, 1) \equiv \text{Unif}(0, 1).
\end{aligned}
\tag{3.1}
$$

Then, the posterior distribution of π_j based on $D_j = (n_j, y_j)$ is given by

$$
\pi_j \mid D_j \sim \text{Beta}(y_j + 1, n_j - y_j + 1), \text{ for } j = 1, \ldots, J.
\tag{3.2}
$$

DOI: 10.1201/9780429052781-3

Unlike continual reassessment method (CRM), which models toxicity across doses using the power model (2.1) or the logistic model, mTPI models toxicity at each dose independently. This is a common feature of model-assisted designs, rendering them computational simplicity and the ability to tabulate all dose escalation/de-escalation rules prior to the trial conduct.

To make the decision of dose escalation/de-escalation, mTPI calculates the unit probability mass (UPM) for each of the three intervals, defined as:

$$\text{UPM1} = \Pr(\pi_j \in (0, \delta_1) \mid D_j) / \delta_1 \,,$$
$$\text{UPM2} = \Pr(\pi_j \in (\delta_1, \delta_2) \mid D_j) / (\delta_2 - \delta_1) \,, \text{and}$$
$$\text{UPM3} = \Pr(\pi_j \in (\delta_2, 1) \mid D_j) / (1 - \delta_2) \,.$$

Suppose j is the current dose level. mTPI determines the next dose as follows:

- If $\text{UPM1} = \max\{\text{UPM1}, \text{UPM2}, \text{UPM3}\}$, then escalate the dose to level $j+1$.

- If $\text{UPM2} = \max\{\text{UPM1}, \text{UPM2}, \text{UPM3}\}$, then stay at the current dose level j.

- If $\text{UPM3} = \max\{\text{UPM1}, \text{UPM2}, \text{UPM3}\}$, then de-escalate the dose to level $j - 1$.

Because the three UPMs can be determined for all possible outcomes $D_j = (n_j, y_j)$, the dose escalation and de-escalation rules can be tabulated before the trial begins, which makes mTPI easy to implement in practice.

This can be done as follows: enumerate all possible values of n_j from 1 up to the maximum sample size N, and given each possible value of n_j, enumerate all possible values of y_j from 0 up to n_j. Then, given a pair of (n_j, y_j), calculate three UPMs and record the resulting dose escalation/de-escalation decision. Table 3.1 shows the dose escalation and de-escalation table for mTPI based on a target DLT rate $\phi = 0.20$ and the proper dosing interval $(\delta_1, \delta_2) = (0.17, 0.23)$.

TABLE 3.1: Decision table of the mTPI design based on a target toxicity rate $\phi = 0.20$ and the proper dosing interval $(\delta_1, \delta_2) = (0.17, 0.23)$, n_j is the number of patients treated at dose level j, and y_j is the number of DLTs observed at dose level j.

Number of patients n_j	1	2	3	4	5	6	7	8	9	10	11	12
Escalate to $j+1$ if $y_j \leq$	0	0	0	0	0	0	0	0	1	1	1	1
De-escalate to $j-1$ if $y_j \geq$	1	1	2	2	3	3	3	4	4	4	5	5
Eliminate levels j to J if $y_j \geq$	NA	2	2	3	3	3	4	4	4	5	5	5

The trial continues until the prespecified maximum sample size N is reached. At that point, the MTD is selected as the dose whose isotonic estimate of the DLT rate is closest to the target ϕ,

$$\text{MTD} = \underset{d_j \in (d_1, \ldots, d_J)}{\arg\min} |\tilde{\pi}_j - \phi|,$$

where $\tilde{\pi}_j$ is the estimate of π_j based on isotonic regression, which can be obtained using the pooled adjacent violators algorithm (Barlow et al., 1972) on $\{\hat{\pi}_j, j = 1, \ldots, J\}$ with $\hat{\pi}_j = y_j/n_j$. Isotonic regression is the technique of fitting a nonparametric or free-form line to observations (e.g., (y_j, n_j)) such that the fitted line is non-decreasing everywhere and lies as close to the observations as possible, see Barlow et al. (1972) for technical details.

For safety, mTPI includes a dose exclusion/safety stopping rule: if $\Pr(\pi_j > \phi \mid D_j) > 0.95$, dose level j and higher are excluded from the trial. If the lowest dose is excluded, the trial is stopped for safety. Such a safety monitoring rule also can be reflected in the decision table (e.g., see the last row of Table 3.1).

One deficiency of mTPI is that its decision rule (i.e., use UPM to guide dose exploration) lacks clear interpretation and leads to a high risk of overdosing patients. To see the problem, consider a trial with a target DLT probability $\phi = 0.20$, and underdosing, proper dosing, and overdosing intervals of $(0, 0.17)$, $(0.17, 0.23)$, and $(0.23, 1)$, respectively. Suppose at a certain stage of the trial, the observed data indicate that the posterior probabilities of the underdosing interval, proper dosing interval, and overdosing interval are 0.01, 0.09, and 0.9, respectively. That is, the observed data indicate that there is a 90% chance that the current dose is overdosing patients and only a 9% chance that the current dose is properly dosing patients. Despite such dominant evidence of overdosing, mTPI does not de-escalate the dose, staying at the same dose for treating the next patient cohort, as UPM1 $= 0.01/0.17 = 0.059$, UPM2 $= 0.09/(0.23 - 0.17) = 1.5$, and UPM3 $= 0.9/(1 - 0.23) = 1.17$.

The pathological behavior of mTPI is caused by UPM non-proportionally weighting the evidence of underdosing, proper dosing, and overdosing due to different lengths of the three intervals. As the overdosing interval is typically the longest, UPM has bias in suppressing the evidence of overdosing. The numerical studies in Section 3.5 confirm this issue.

3.3 Keyboard design

Yan et al. (2017) proposed the keyboard design as a seamless upgrade of the mTPI design to address the latter's overdosing issue. Unlike the mTPI design, which divides the toxicity probabilities into three intervals (i.e., underdosing, proper dosing and overdosing), the keyboard design defines a series of equal-width dosing intervals (referred to as "keys") that correspond to all potential locations of the true toxicity rate of a particular dose. The design then uses the key with the highest posterior probability (referred to as the "strongest key")

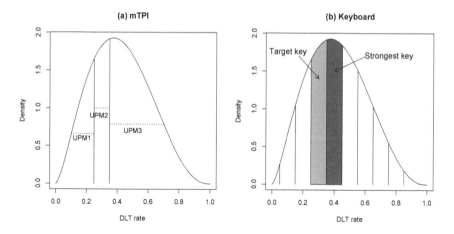

FIGURE 3.1: Contrast between (a) the mTPI design and (b) the keyboard design. The curves are the posterior distributions of π_j. To determine the next dose, the mTPI design compares the values of the three UPMs, whereas the keyboard design compares the location of the strongest key with respect to the target key.

to guide dose escalation and de-escalation. Figure 3.1 contrasts the keyboard and mTPI designs.

The keyboard design starts by specifying a proper dosing interval $\mathcal{I}^* = (\delta_1, \delta_2)$, referred to as the "target key," and then populates this interval toward both sides of the target key, forming a series of keys of an equal width. The keys span the range of 0 to 1. For example, given the proper dosing interval or target key of (0.25, 0.35), on its left side we form two keys of width 0.1 (i.e., (0.15, 0.25) and (0.05, 0.15)); and on its right side we form six keys of width 0.1 (i.e., (0.35, 0.45), (0.45, 0.55), (0.55, 0.65), (0.65, 0.75), (0.75, 0.85), and (0.85, 0.95)). We denote the resulting intervals/keys as $\mathcal{I}_1, \cdots, \mathcal{I}_K$. As all keys must have an equal width and be within [0, 1], some DLT probability values at the two ends (e.g., < 0.05 or > 0.95 in the example) may not be covered by keys. As explained in Yan et al. (2017), ignoring these "residual" DLT rates at the two ends does not pose any issue for the decision making of dose escalation and de-escalation. This is because the posterior distribution of π_j is unimodal, and the decision is made by the relative position of the target key to the strongest key.

To make the decision of dose escalation and de-escalation, given the observed data $D_j = (n_j, y_j)$ at the current dose level j, the keyboard design identifies the interval $\mathcal{I}_{\max}$ that has the largest posterior probability,

$$\mathcal{I}_{\max} = \underset{\mathcal{I}_1, \cdots, \mathcal{I}_K}{\arg\max}\{\Pr(\pi_j \in \mathcal{I}_k \mid D_j)\},$$

which can be easily evaluated based on π_j's posterior distribution given by equation (3.2), assuming that π_j follows a beta-binomial model (3.1). $\mathcal{I}_{\max}$ represents the interval that the true value of π_j is most likely located, referred to as the "strongest key." Graphically, the strongest key is the one with the largest area under the posterior distribution curve of π_j, see panel (b) of Figure 3.1. If the strongest key is on the left (or right) side of the target key, then it means the observed data suggest that the current dose is most likely to represent underdosing (or overdosing), and thus dose escalation (or de-escalation) is needed. If the strongest key is the target key, the observed data support that the current dose is most likely to be in the proper dosing interval, and thus it is desirable to stay at the current dose for treating the next patient. In contrast, UPM used by the mTPI design does not have such an intuitive interpretation and tends to distort the evidence of overdosing, as described previously.

Based on the keyboard design, patients in the first cohort are treated at the lowest dose level, or the physician-specified level. Suppose j is the current dose level. The keyboard design determines the next dose as follows:

- If the strongest key $\mathcal{I}_{\max}$ is on the left side of the target key $\mathcal{I}^*$, then escalate the dose to level $j + 1$.

- If $\mathcal{I}_{\max}$ is $\mathcal{I}^*$, then stay at the current dose level j.

- If $\mathcal{I}_{\max}$ is on the right side of $\mathcal{I}^*$, then de-escalate the dose to level $j - 1$.

The trial continues until the maximum sample size N is reached. At that point, the MTD is selected based on isotonic estimates $\tilde{\pi}_j$ as described previously.

During the trial conduct, the keyboard design imposes the dose exclusion/early stopping rule such that if $\Pr(\pi_j > \phi \mid D_j) > 0.95$ and $n_j \geq 3$, dose level j and higher are eliminated from the trial, and the trial is terminated if the lowest dose is eliminated, where $\Pr(\pi_j > \phi \mid D_j)$ is evaluated based on the posterior distribution (3.2).

Similar to the mTPI design, the dose escalation and de-escalation rules of the keyboard design can be tabulated for each possible $n_j = 1, \cdots, N$ before the trial begins, making it easy to implement in practice. For example, Table 3.2 shows the dose escalation and de-escalation boundaries of the keyboard design based on a target toxicity rate $\phi = 0.20$ and the proper dosing interval $(\delta_1, \delta_2) = (0.17, 0.23)$. By contrasting Tables 3.1 and 3.2, we find that given the same proper dosing interval the keyboard design is less aggressive than the mTPI design. In particular, when three out of eight patients have toxicities at a given dose level, the keyboard design recommends dose-escalation, whereas the mTPI design suggests retaining the dose even though this dose yields an observed DLT rate of $3/8 = 37.5\%$ with $\Pr(\pi_j > 0.20 \mid n_j = 8, y_j = 3) = 0.91$.

As the location of the strongest key approximately indicates the mode of the posterior distribution of π_j, the keyboard design can be viewed as a posterior mode-based dose-finding method. Yan et al. (2017) shows that the keyboard design substantially outperforms mTPI. Of note, the variation of

TABLE 3.2: Decision table of the keyboard design based on a target toxicity rate $\phi = 0.20$ and the proper dosing interval $(\delta_1, \delta_2) = (0.17, 0.23)$, n_j is the number of patients treated at dose level j, and y_j is the number of DLTs observed at dose level j.

Number of patients n_j	1	2	3	4	5	6	7	8	9	10	11	12
Escalate to $j+1$ if $y_j \leq$	0	0	0	0	0	1	1	1	1	1	1	2
De-escalate to $j-1$ if $y_j \geq$	1	1	1	1	2	2	2	2	3	3	3	3
Eliminate levels j to J if $y_j \geq$	NA	NA	2	3	3	3	4	4	4	5	5	5

mTPI, mTPI-2 (Guo and Yuan, 2017), is equivalent to the keyboard design, but is perplexing and less transparent than the keyboard design. mTPI-2 relies on complicated procedures, such as Occam's razor and model selection, which are difficult to understand and communicate with non-statisticians.

3.4   Bayesian optimal interval (BOIN) design

3.4.1   Trial design

Unlike the mTPI and keyboard designs, which require specifying the prior distribution of π_j and calculating π_j's posterior for each of the possible values of (n_j, y_j), the BOIN design is more transparent and straightforward (Yuan et al., 2016a). Under the BOIN design, the decision of dose escalation and de-escalation involves only a simple comparison of the observed DLT rate at the current dose with a pair of fixed, prespecified dose escalation and de-escalation boundaries (Liu and Yuan, 2015; Yuan et al., 2016a). In addition, it has a theoretical foundation of possessing the optimal property by minimizing the decision error. Due to its transpancy and competitive and robust performance, the US Food and Drug Administration (FDA) granted the BOIN design the fit-for-purpose designation as a drug development tool in 2021 (https://www.fda.gov/drugs/development-approval-process-drugs/drug-development-tools-fit-purpose-initiative). This is the first MTD-finding design that has received this designation.

Specifically, let $\hat{\pi}_j = y_j/n_j$ denote the observed DLT rate at the current dose, and λ_e and λ_d denote the optimal dose escalation and de-escalation boundaries, respectively, which will be described later. The BOIN design is illustrated in Figure 3.2 and described as follows:

1. Patients in the first cohort are treated at the lowest dose d_1, or the physician-specified dose.

TABLE 3.3: The escalation/de-escalation boundaries (λ_e, λ_d) under the BOIN design for different target toxicity rates ϕ, using the default underdosing toxicity probability $\phi_1 = 0.6\phi$ and overdosing toxicity probability $\phi_2 = 1.4\phi$.

| Boundaries | Target toxicity probability ϕ | | | | | |
	0.15	0.2	0.25	0.3	0.35	0.4
λ_e	0.118	0.157	0.197	0.236	0.276	0.316
λ_d	0.179	0.238	0.298	0.358	0.419	0.479

2. Suppose j is the current dose level; to assign a dose to the next cohort of patients:

   - if $\hat{\pi}_j \leq \lambda_e$, escalate the dose to level $j + 1$.

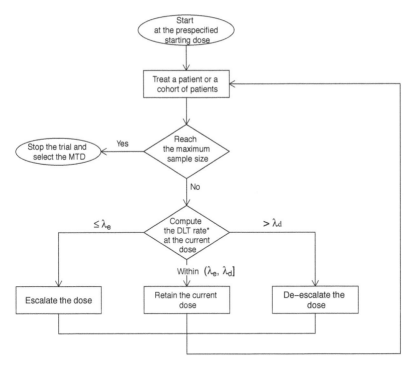

FIGURE 3.2: BOIN design flowchart.

- if $\hat{\pi}_j > \lambda_d$, de-escalate the dose to level $j-1$.
- otherwise, (i.e., $\lambda_e < \hat{\pi}_j \leq \lambda_d$), stay at the current dose level j.

3. Repeat step 2 until the maximum sample size N is reached. At that point, select MTD as the dose whose isotonic estimate $\tilde{\pi}_j$ is closest to the target ϕ, where $\tilde{\pi}_j$ is obtained by applying the pooled adjacent violators algorithm (Barlow et al., 1972) to $\{\hat{\pi}_j, j = 1, \ldots, J\}$.

During the trial conduct, the BOIN design imposes a dose elimination (or overdose control) rule as follows:

if $\Pr(\pi_j > \phi \,|\, D_j) > 0.95$ and $n_j \geq 3$, dose levels j and higher are eliminated from the trial, and the trial is terminated if the lowest dose level is eliminated.

The value of $\Pr(\pi_j > \phi \,|\, D_j)$ is evaluated based on the posterior distribution (3.2). When the trial is terminated due to toxicity, no dose should be selected as MTD.

Table 3.3 provides the optimal dose escalation and de-escalation boundaries (λ_e, λ_d) for commonly-used target DLT probabilities ϕ. For example, given $\phi = 0.3$, the corresponding escalation boundary $\lambda_e = 0.236$ and the de-escalation boundary $\lambda_d = 0.358$. That is, escalate/de-escalate/retain the dose if $0/3$ or $2/3$ or $1/3$ patients have DLT, respectively, given that three patients have been treated at the current dose.

Table 3.4, or equivalently Figure 3.3, shows the discretized escalation/de-escalation boundaries up to $n_j = 12$ when $\phi = 0.3$. This discretized version of the decision table is often handy in practice to conduct the trial, which can be easily generated using the software described later.

The hallmark of the BOIN design is its simplicity and transparency. Arguably, its decision rule is even simpler than the 3+3 design; it just involves a simple comparison between $\hat{\pi}_j$ and (λ_e, λ_d). Such simplicity renders BOIN several important advantages over other model-assisted designs (e.g., the mTPI and keyboard designs). From the statistical viewpoint, $\hat{\pi}_j$ is the (nonparametric) maximum likelihood estimate of π_j, and it enjoys desirable statistical properties such as being consistent and efficient. From a practical viewpoint, $\hat{\pi}_j$ is the most natural and intuitive estimate of π_j that is accessible by non-statisticians, making the BOIN design simpler and more transparent than the mTPI and keyboard designs. In our experience, explaining the BOIN design to clinicians, especially when equipped with the flowchart displayed in Figure 3.2, is easy and well received.

In addition, due to the feature that the BOIN design guarantees de-escalating the dose when $\hat{\pi}_j > \lambda_d$, it is easy for clinicians and regulatory agencies to assess the safety of a trial using the BOIN design. For example, given a target DLT rate $\phi = 0.25$, we know *a priori* that a phase I trial using the BOIN design guarantees de-escalating the dose if the observed DLT rate is

TABLE 3.4: The escalation/de-escalation boundaries of the BOIN design up to 12 patients at a dose when the target toxicity rate $\phi = 0.3$.

	The number of patients treated at the current dose											
Action	1	2	3	4	5	6	7	8	9	10	11	12
Escalate if no. of DLT $<=$	0	0	0	0	1	1	1	1	2	2	2	2
De-escalate if no. of DLT $>=$	1	1	2	2	2	3	3	3	4	4	4	5
Eliminate if no. of DLT $>=$	NA	NA	3	3	4	4	5	5	5	6	6	7

"no. of DLT" is the number of patients with at least one DLT. When none of the actions (i.e., escalate, de-escalate, or eliminate) is triggered, stay at the current dose for treating the next cohort of patients. "NA" means that a dose cannot be eliminated before treating three evaluable patients.

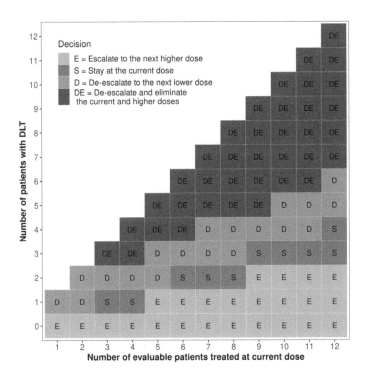

FIGURE 3.3: BOIN design decision table up to 12 patients at a dose when the target toxicity rate $\phi = 0.3$.

higher than $\lambda_d = 0.298$ (i.e., the default de-escalation boundary). Accordingly, the BOIN design also allows users to easily calibrate the design to satisfy a specific safety requirement mandated by regulatory agencies by choosing an appropriate target DLT rate ϕ. For example, consider a phase I trial with a new compound for which the regulatory agency mandates a dose de-escalation for any observed toxicity rate higher than 0.25. We can easily fulfill that requirement by setting the target DLT rate $\phi = 0.21$ (based on equation (3.9) described in the next section), under which BOIN automatically guarantees de-escalating the dose if the observed toxicity rate $\hat{\pi}_j > \lambda_d = 0.25$. Such flexibility and transparency grants the BOIN design an important advantage in practice.

Another unique strength of BOIN is its Bayesian-frequentist dual interpretation, making the design appealing to wider audiences. In contrast, the mTPI and keyboard designs only have a Bayesian interpretation and require specifying priors and calculating posterior distributions. As shown in the following sections, BOIN is derived as a Bayesian optimal design, but has frequentist interpretation and optimality (i.e., under the non-informative prior, the BOIN decision rule is equivalent to using the likelihood ratio test to determine dose escalation/de-escalation (Liu and Yuan, 2015)). That is also why the BOIN's decision rules (with the non-informative prior) have the appearance of a classical frequentist design and only involve the observed DLT rate $\hat{\pi}_j$, the maximum likelihood estimate of π_j.

Despite its simplicity, numerical studies show that BOIN yields excellent performance comparable to the model-based CRM design (Zhou et al., 2018a,b; Ruppert and Shoben, 2018). BOIN outperforms mTPI with higher accuracy for identifying MTD and a substantially lower risk of overdosing patients, and it is simpler and more transparent than the keyboard (or mTPI-2) design. In addition, BOIN also is more versatile; it can handle drug-combination trials (Lin and Yin, 2017a; Zhang and Yuan, 2016), late-onset toxicity (Yuan et al., 2018), low-grade toxicities (Lin, 2018), and toxicity and efficacy jointly (Takeda et al., 2018; Zhou et al., 2019; Lin et al., 2020b), and incorporate historical information and real-world evidence (Zhou et al., 2021a). These extensions of BOIN will be covered in later chapters.

More importantly, BOIN has been widely validated in practice. It has been used in variety of oncology trials, including trials for pediatric tumors, adult tumors, solid tumors, and liquid tumors, see Yuan et al. (2019) for a list of completed and ongoing trials that have implemented the BOIN design. Additionally, BOIN has been used in non-oncology trials such as stem cell therapy for patients with stroke (Phan et al., 2018).

3.4.2 Theoretical derivation

This section describes the theoretical foundation of the BOIN design, which provides the basis for various extensions of the design (e.g., drug combination, late-onset toxicity, and toxicity-efficacy tradeoff) described in later chapters.

The BOIN design is derived based on the optimal design theory. Let $\lambda_e(d_j, n_j, \phi)$ and $\lambda_d(d_j, n_j, \phi)$ denote the general dose escalation and de-escalation boundaries that are unspecified functions of the dose (i.e., d_j), the number of patients treated (i.e., n_j), and the DLT target (i.e., ϕ), where $0 \leq \lambda_e(d_j, n_j, \phi) < \lambda_d(d_j, n_j, \phi) < 1$. Consider a class of nonparametric designs $\mathcal{C}_{np}$:

1. Patients in the first cohort are treated at the lowest dose d_1, or the physician-specified dose.

2. Suppose j is the current dose level; to assign a dose to the next cohort of patients:

   - if $\hat{\pi}_j \leq \lambda_e(d_j, n_j, \phi)$, escalate the dose level to $j + 1$.
   - if $\hat{\pi}_j > \lambda_d(d_j, n_j, \phi)$, de-escalate the dose level to $j - 1$.
   - otherwise, i.e., $\lambda_e(d_j, n_j, \phi) < \hat{\pi}_j \leq \lambda_d(d_j, n_j, \phi)$, stay at the current dose level j.

3. Repeat step 2 until the maximum sample size N is reached.

As escalation and de-escalation boundaries $\lambda_e(d_j, n_j, \phi)$ and $\lambda_d(d_j, n_j, \phi)$ can freely vary according to d_j, n_j, and ϕ, this class of designs are extremely broad and contain all possible nonparametric designs that do not impose the parametric assumption on the dose-toxicity curve and make dose escalation and de-escalation based on the local data $D_j = (y_j, n_j)$. The mTPI, keyboard/mTPI-2, and BOIN designs all belong to $\mathcal{C}_{np}$. For notational brevity, we use shorthands $\lambda_{ej} \equiv \lambda_e(d_j, n_j, \phi)$ and $\lambda_{dj} \equiv \lambda_d(d_j, n_j, \phi)$.

The BOIN design is obtained by minimizing the probability of making incorrect decisions of dose escalation and de-escalation within $\mathcal{C}_{np}$. Liu and Yuan (2015) described two versions of the BOIN design: the local optimal BOIN design, which is optimized based on point hypotheses, and the global optimal BOIN design, which is optimized based on interval hypotheses. Liu and Yuan (2015) recommended the local optimal BOIN design because of its better performance. Thus, we here focus on the development of the local optimal BOIN design and simply call it the BOIN design.

To proceed, define three point hypotheses:

$$H_{0j} : \pi_j = \phi, \qquad H_{1j} : \pi_j = \phi_1, \qquad H_{2j} : \pi_j = \phi_2,$$

where ϕ_1 indicates that the dose is substantially underdosing (i.e., below MTD) such that escalation is required, and ϕ_2 indicates that the dose is substantially overdosing such that de-escalation is required. Let $\mathcal{S}$, $\mathcal{E}$, and $\mathcal{D}$ denote the decisions of stay (at the current dose), escalate, and de-escalate, respectively. Under H_{0j}, the correct decision is $\mathcal{S}$, and incorrect decisions are $\bar{\mathcal{S}} = \{\mathcal{E}, \mathcal{D}\}$; under H_{1j}, the correct decision is $\mathcal{E}$, and incorrect decisions are

$\bar{\mathcal{E}} = \{\mathcal{S}, \mathcal{D}\}$; and under H_{2j}, the correct decision is $\mathcal{D}$, and incorrect decisions are $\bar{\mathcal{D}} = \{\mathcal{S}, \mathcal{E}\}$.

Under the Bayesian paradigm, we assign each of the hypotheses a prior probability of being true, denoted as $\omega_{kj} = \Pr(H_{kj}), k = 0, 1, 2$. The probability of making an incorrect decision (the decision error rate), denoted as $\alpha(\lambda_{ej}, \lambda_{dj})$, at each of the dose assignments is given by

$$
\begin{aligned}
&\alpha(\lambda_{ej}, \lambda_{dj}) \\
&= \Pr(H_{0j}) \Pr(\bar{\mathcal{S}}|H_{0j}) + \Pr(H_{1j}) \Pr(\bar{\mathcal{E}}|H_{1j}) + \Pr(H_{2j}) \Pr(\bar{\mathcal{D}}|H_{2j}) \\
&= \Pr(H_{0j}) \Pr(y_j \le n_j \lambda_{ej} \text{ or } y_j > n_j \lambda_{dj}|H_{0j}) + \Pr(H_{1j}) \Pr(y_j > n_j \lambda_{ej}|H_{1j}) \\
&\quad + \Pr(H_{2j}) \Pr(y_j \le n_j \lambda_{dj}|H_{2j}) \\
&= \omega_{0j}\{Bin(n_j\lambda_{ej}; n_j, \phi) + 1 - Bin(n_j\lambda_{dj}; n_j, \phi)\} + \\
&\quad \omega_{1j}\{1 - Bin(n_j\lambda_{ej}; n_j, \phi_1)\} + \omega_{2j}Bin(n_j\lambda_{dj}; n_j, \phi_2), \quad (3.3)
\end{aligned}
$$

where $Bin(b; n, \phi)$ is the cumulative density function (CDF) of the binomial distribution, with size and probability parameters n and ϕ evaluated at the value b. We rewrite the decision error $\alpha(\lambda_{ej}, \lambda_{dj})$ as

$$
\alpha(\lambda_{ej}, \lambda_{dj}) = \alpha_1(\lambda_{ej}) + \alpha_2(\lambda_{dj}) + \omega_{0j} + \omega_{1j},
$$

where

$$
\begin{aligned}
\alpha_1(\lambda_{ej}) &= \omega_{0j}Bin(n_j\lambda_{ej}; n_j, \phi) - \omega_{1j}Bin(n_j\lambda_{ej}; n_j, \phi_1) \\
\alpha_2(\lambda_{dj}) &= \omega_{2j}Bin(n_j\lambda_{dj}; n_j, \phi_2) - \omega_{0j}Bin(n_j\lambda_{dj}; n_j, \phi).
\end{aligned}
$$

To minimize $\alpha(\lambda_{ej}, \lambda_{dj})$, we can minimize $\alpha_1(\lambda_{ej})$ and $\alpha_2(\lambda_{dj})$ separately with regard to λ_{ej} and λ_{dj}, respectively. As $\alpha_1(\lambda_{ej})$ and $\alpha_2(\lambda_{dj})$ are symmetric, below we consider the minimization of $\alpha_1(\lambda_{ej})$:

$$
\begin{aligned}
\alpha_1(\lambda_{ej}) &= \omega_{0j}Bin(n_j\lambda_{ej}; n_j, \phi) - \omega_{1j}Bin(n_j\lambda_{ej}; n_j, \phi_1) \\
&= \sum_{y=0}^{\lfloor n_j\lambda_{ej} \rfloor} \binom{n_j}{y} \{\omega_{0j}\phi^y(1-\phi)^{n_j-y} - \omega_{1j}\phi_1^y(1-\phi_1)^{n_j-y}\} \\
&= \sum_{y=0}^{\lfloor n_j\lambda_{ej} \rfloor} \omega_{1j}\binom{n_j}{y}\phi_1^y(1-\phi_1)^{n_j-y} \left\{ \frac{\omega_{0j}}{\omega_{1j}}\left(\frac{\phi}{\phi_1}\right)^y\left(\frac{1-\phi}{1-\phi_1}\right)^{n_j-y} - 1 \right\}.
\end{aligned}
$$

By the definition of the CDF, $\alpha_e(\lambda_{ej}) = 0$ when $\lfloor n_j\lambda_{ej} \rfloor < 0$ and $\alpha_1(\lambda_{ej}) = 1$ when $\lfloor n_j\lambda_{ej} \rfloor \ge n_j$.

Assuming that y^* is continuous and setting

$$
\frac{\omega_{0j}}{\omega_{1j}}\left(\frac{\phi}{\phi_1}\right)^{y^*}\left(\frac{1-\phi}{1-\phi_1}\right)^{n_j-y^*} - 1 = 0, \quad (3.4)
$$

we obtain

$$
y^* = \frac{n_j\log\left(\dfrac{1-\phi_1}{1-\phi}\right) + \log\left(\dfrac{\omega_{1j}}{\omega_{0j}}\right)}{\log\left(\dfrac{\phi(1-\phi_1)}{\phi_1(1-\phi)}\right)}.
$$

Because $\phi > \phi_1$, $\left(\frac{\phi}{\phi_1}\right)^y \left(\frac{1-\phi}{1-\phi_1}\right)^{n_j-y}$ monotonically increases with y. It follows that $\left\{\frac{\pi_{0j}}{\pi_{1j}}\left(\frac{\phi}{\phi_1}\right)^y \left(\frac{1-\phi}{1-\phi_1}\right)^{n_j-y} - 1\right\} \geq 0$ when $y \geq y^*$, and < 0 when $y < y^*$. Therefore, given $0 \leq y \leq n_j$, $\alpha_1(\lambda_{ej})$ is minimized when

$$
n_j \lambda_{ej} \in \begin{cases}
[n_j - I(y^* = n_j), \infty), & \text{if} \quad y^* \geq n_j, \\
[\lceil y^* \rceil - 1, \lfloor y^* \rfloor + 1), & \text{if} \quad 0 < y^* < n_j, \\
(-\infty, I(y^* = 0)), & \text{if} \quad y^* \leq 0,
\end{cases}
$$

where $I(\cdot)$ is an indicator function. This leads to the solution

$$
\lambda_{ej} \in \begin{cases}
\left[1 - \dfrac{I(y^* = n_j)}{n_j}, \infty\right), & \text{if} \quad y^* \geq n_j, \\[2mm]
\left[\dfrac{\lceil y^* \rceil - 1}{n_j}, \dfrac{\lfloor y^* \rfloor + 1}{n_j}\right), & \text{if} \quad 0 < y^* < n_j, \\[2mm]
\left(-\infty, \dfrac{I(y^* = 0)}{n_j}\right), & \text{if} \quad y^* \leq 0.
\end{cases}
$$
(3.5)

Because of the symmetry of $\alpha_1(\lambda_{ej})$ and $\alpha_2(\lambda_{dj})$, it follows that the solution that minimizes $\alpha_2(\lambda_{dj})$ is given by

$$
\lambda_{dj} \in \begin{cases}
\left[1 - \dfrac{I(y^{**} = n_j)}{n_j}, \infty\right), & \text{if} \quad y^{**} \geq n_j, \\[2mm]
\left[\dfrac{\lceil y^{**} \rceil - 1}{n_j}, \dfrac{\lfloor y^{**} \rfloor + 1}{n_j}\right), & \text{if} \quad 0 < y^{**} < n_j, \\[2mm]
\left(-\infty, \dfrac{I(y^{**} = 0)}{n_j}\right), & \text{if} \quad y^{**} \leq 0,
\end{cases}
$$
(3.6)

where

$$
y^{**} = \frac{n_j \log\left(\dfrac{1-\phi}{1-\phi_2}\right) + \log\left(\dfrac{\omega_{0j}}{\omega_{2j}}\right)}{\log\left(\dfrac{\phi_2(1-\phi)}{\phi(1-\phi_2)}\right)}.
$$

As any values of λ_{ej} and λ_{dj} located in the interval solutions (3.5) and (3.6) produce the same error rate, for the purpose of designing the trial and making decisions, a point solution located in the interval solutions is sufficient. As $\lceil x \rceil - 1 < x < \lfloor x \rfloor + 1$, one specific "middle" point solution is

$$
\lambda_{ej}^* = y^*/n_j = \frac{\log\left(\dfrac{1-\phi_1}{1-\phi}\right) + n_j^{-1}\log\left(\dfrac{\omega_{1j}}{\omega_{0j}}\right)}{\log\left(\dfrac{\phi(1-\phi_1)}{\phi_1(1-\phi)}\right)},
$$
(3.7)

$$\lambda_{dj}^* = y^{**}/n_j = \frac{\log\left(\frac{1-\phi}{1-\phi_2}\right) + n_j^{-1}\log\left(\frac{\omega_{0j}}{\omega_{2j}}\right)}{\log\left(\frac{\phi_2(1-\phi)}{\phi(1-\phi_2)}\right)}. \tag{3.8}$$

This is the solution provided by Liu and Yuan (2015). The above derivation provides a more complete interval solution pair (3.5) and (3.6).

When the non-informative prior $\omega_{0j} = \omega_{1j} = \omega_{2j} = 1/3$ is used, which is recommended for most trials, the optimal escalation boundaries become

$$\lambda_e = \frac{\log\left(\frac{1-\phi_1}{1-\phi}\right)}{\log\left(\frac{\phi(1-\phi_1)}{\phi_1(1-\phi)}\right)}, \qquad \lambda_d = \frac{\log\left(\frac{1-\phi}{1-\phi_2}\right)}{\log\left(\frac{\phi_2(1-\phi)}{\phi(1-\phi_2)}\right)}, \tag{3.9}$$

which are independent of j and n_j. This results in the hallmark simplicity of BOIN — using a pair of fixed dose escalation and de-escalation boundaries throughout the trial to guide the dose transition.

As noted by Liu and Yuan (2022), the original publication of the BOIN design (Liu and Yuan, 2015) had incorrectly specified the de-escalation rule as: de-escalate the dose if $\hat{\pi}_j \geq \lambda_d$, which should be $\hat{\pi}_j > \lambda_d$, as shown above. That error does not affect the application of the BOIN design as it is virtually impossible that $\hat{\pi}_j = \lambda_d$ in practice. Liu and Yuan (2022) examined the target DLT rate $\phi \in \{0.1000, 0.1001, 0.1002, \ldots, 0.4000\}$ with $n_j = 30$ (i.e., up to treating 30 patients per dose), and did not find any case that $\hat{\pi}_j = \lambda_d$. In addition, de-escalation if $\hat{\pi}_j \geq \lambda_d$ is slightly more conservative than de-escalation if $\hat{\pi}_j > \lambda_d$. Therefore, even if the equal sign could be taken, there are no safety concerns that arise for the trials using the "original" rule.

3.4.3   Specification of design parameters

The specification of ϕ_1 and ϕ_2 are critical as they determine the dose escalation and de-escalation boundaries (λ_e, λ_d). Liu and Yuan (2015) recommended the default values of $\phi_1 = 0.6\phi$ and $\phi_2 = 1.4\phi$ for general use, which yield good operating characteristics as confirmed by numerous subsequent simulation studies and real clinical trials. These are the default values used in the BOIN software described later. Therefore, the BOIN design is essentially calibration-free: once the target DLT rate ϕ is specified, the design (i.e., escalation and de-escalation boundaries λ_e and λ_d) is determined and ready to use.

When needed, however, the values of ϕ_1 and ϕ_2 can be tuned to achieve a particular requirement of the trial at hand. For example, if more conservative dose escalation is required, we may set $\phi_2 = 1.2\phi$. The performance of BOIN is generally robust to the specification of ϕ_1 and ϕ_2, but we should refrain from setting ϕ_1 and ϕ_2 too close to ϕ (e.g., a difference less than 20% of ϕ). This is because the small sample size of phase I trials provides virtually no power to distinguish a difference smaller than that, and minimizing decision errors under such hypotheses does not make any practical sense. For example, given the sample size of 30 patients per dose, based on Fisher's exact test, we only have 7% power to distinguish a dose with a toxicity rate of 0.3 from the other with toxicity rate of 0.35, at the one-sided significance level of 0.05, using R function `power.fisher.test( )` in the `statmod` package.

As a side note, ϕ_1 and ϕ_2 used in the BOIN design have different interpretations than the proper dosing interval (δ_1, δ_2) used in the mTPI and keyboard designs. Specifically, ϕ_1 and ϕ_2 represent the DLT rates that are unacceptable (i.e., underdosing or overdosing such that dose escalation or de-escalation is required); whereas δ_1 and δ_2 represent the range of DLT probabilities that are acceptable. For example, given that the target DLT probability $\phi = 0.25$, setting $\phi_1 = 0.15$ and $\phi_2 = 0.35$ means that the doses with the DLT rates of 0.15 and 0.35 are regarded as unacceptably underdosing and overdosing, respectively, whereas setting $\delta_1 = 0.15$ and $\delta_2 = 0.35$ means that the dose with a DLT rate between 0.15 and 0.35 is regarded as acceptable. Thus, in general, the value of ϕ_1 should be smaller than that of δ_1 and the value of ϕ_2 should be greater than that of δ_2.

3.4.4   Statistical properties

Bayesian and frequentist optimality

One unique property of BOIN is that it enjoys both Bayesian and frequentist interpretation and optimality.

Theorem 3.1 *The dose escalation and de-escalation boundaries (λ_e, λ_d) of the BOIN design are the boundaries from where $\Pr(H_1|D_j) \geq \Pr(H_0|D_j)$ and $\Pr(H_2|D_j) > \Pr(H_0|D_j)$, respectively. When the non-informative prior is used, the boundaries are the likelihood ratio test boundaries.*

In other words, the decision rule of BOIN is equivalent to the following intuitive Bayesian decision rule:

- If $\dfrac{\Pr(H_{1j}|D_j)}{\Pr(H_{0j}|D_j)} \geq 1$ (i.e., the data indicate that H_{1j} is equal or more likely to be true than H_{0j}), escalate the dose.

- If $\dfrac{\Pr(H_{2j}|D_j)}{\Pr(H_{0j}|D_j)} > 1$ (i.e., the data indicate that H_{2j} is more likely to be true than H_{0j}), de-escalate the dose.

The proof of Theorem 3.1 is straightforward noting that the ratio in equation (3.4) is the $\Pr(H_{1j}|D_j)/\Pr(H_{0j}|D_j)$.

Define $L(D_j|H_{0j}) \propto \phi^{y_j}(1 - \phi)^{(n_j - y_j)}$ as the binomial likelihood function of the data $D_j = (n_j, y_j)$ under H_{0j}, and similarly define $L(D_j|H_{kj})$ as the likelihood function under $H_{kj}, k = 1, 2$. When the non-informative prior is used, the decision rule of BOIN is equivalent to the following frequentist decision rule:

- If $\dfrac{L(D_j|H_{1j})}{L(D_j|H_{0j})} \geq 1$ (i.e., the likelihood of D_j under H_{1j} is equal to or greater than that under H_{0j}), escalate the dose.

- If $\dfrac{L(D_j|H_{2j})}{L(D_j|H_{0j})} > 1$ (i.e., the likelihood of D_j under H_{1j} is greater than that under H_{0j}), de-escalate the dose.

Long-memory coherence

Coherence is a finite-sample property that describes how a phase I design behaves in dose escalation and de-escalation in light of observed DLT data. Cheung (2005) originally defined coherence as a design property by which dose escalation (or de-escalation) is prohibited when the most recently treated patient experiences (or does not experience) toxicity. Liu and Yuan (2015) extended that concept and defined two different types of coherence: short-memory coherence and long-memory coherence. They referred to the coherence proposed by Cheung (2005) as short-memory coherence because it concerns the observation from only the most recently treated patient, ignoring the observations from the patients who were previously treated. By contrast, long-memory coherence concerns the accumulated data observed from the most recent dose level.

DEFINITION (Short-memory coherence) A design is called short-memory coherent if it never escalates the dose when the most recently treated patient experiences DLT, and never de-escalates the dose when the most recently treated patient does not experience DLT.

DEFINITION (Long-memory coherence) A design is called long-memory coherent if it never escalates the dose when the observed DLT rate at the current dose is higher than the target ϕ, and never de-escalates the dose when the observed DLT rate at the current dose is lower than ϕ.

From a practical viewpoint, long-memory coherence is more relevant because when clinicians determine whether a dose escalation/de-escalation is practically plausible, they almost always base their decision on the toxicity data from all patients treated at the current dose, rather than only the single patient most recently treated. This is more important considering that, patients in phase I trials are highly heterogeneous, and the toxicity outcome from a single patient can be spurious. For example, suppose the target DLT rate $\phi = 0.3$ and, at the current dose, the most recently treated patient experienced DLT but none of the nine patients previously treated at the same dose had DLT. As the overall observed DLT rate at the current dose is 1/10, escalating the dose should not be regarded as an inappropriate action, although it violates short-memory coherence.

Theorem 3.2 *The BOIN design based on the non-informative prior with* $\omega_{0j} = \omega_{1j} = \omega_{2j} = 1/3$ *is long-memory coherent.*

The proof of Theorem 3.2 is straightforward based on the equation (3.9). Besides the above finite-sample properties, Liu and Yuan (2015) showed that, assuming the existence of the target dose (i.e., at least a dose located in (λ_e, λ_d)), the BOIN design also has the following desirable large-sample property.

Theorem 3.3 *As the number of patients goes to infinity, the dose assignment and the selection of MTD under the BOIN design converge almost surely to dose level j^* if dose level j^* is the only dose satisfying $\pi_{j^*} \in (\lambda_e, \lambda_d)$. If there are multiple dose levels in (λ_e, λ_d), the design will converge almost surely to one of these levels.*

As a side note, one might be concerned that BOIN converges to the "stay" interval (λ_e, λ_d), rather than the target ϕ. Actually, this is not a concern. As noted by Zhou et al. (2021b), under large samples it is more appropriate to use the local alternative hypotheses $H_1 : \pi_j = \phi - \Delta_1/\sqrt{n}$ and $H_2 : \pi_j = \phi + \Delta_2/\sqrt{n}$, where Δ_1 and Δ_2 are constant, rather than fixed alternatives $H_1 : \pi_j = \phi_1$ and $H_2 : \pi_j = \phi_2$ as used in the above theorem. Then, as (λ_e, λ_d) converges to ϕ, the dose assignment and the selection of MTD under the BOIN design naturally converge almost surely to ϕ.

3.4.5 Frequently asked questions

In this section, we describe several frequently asked questions with answers to provide more insights on BOIN and model-assisted designs.

1. Does the BOIN decision rule account for the variance of $\hat{\pi}_j$ (or equivalently, the sample size n_j)?

The answer is yes. The simplicity of the BOIN design (i.e., making decisions by comparing $\hat{\pi}_j$ with λ_e and λ_d) might lead one to think that the BOIN decision rule does not consider the variance of $\hat{\pi}_j$ (or equivalently, the sample size n_j). This, however, is not true. As shown by equation (3.3), the derivation and minimization of the decision error α depends on the sampling distribution of $\hat{\pi}_j$, thus it directly accounts for the uncertainty of $\hat{\pi}_j$. The optimal decision boundaries independent of n_j should not be mistakenly regarded as ignoring n_j.

To help readers to understand this point, consider an experiment of drawing balls with replacement from a bag of red and black balls. There are a total of 9 balls in the bag, but we do not know if there are more red or black balls. The objective is to determine if there are more red or black balls. The experiment is to randomly draw a ball from the bag, record the color, put it back, and repeat. Clearly, no matter whether we do the experiment 3 or 30 times, as long as we see more red balls, the best decision is to claim that there are more red balls. The only difference is that the decision based on 30 experiments has a smaller decision error, although both minimize the decision error. This is exactly how the Bayes classifier works, which optimizes and minimizes the Bayes error rate (Berger, 2013).

2. Is it reasonable to assume the non-informative prior $\omega_{0j} = \omega_{1j} = \omega_{2j} = 1/3$ for all doses $d_1 < \cdots < d_J$?

The answer is yes. This seems counterintuitive: as toxicity monotonically increases with the dose, it seems natural to assume that high doses have higher prior probabilities of overdosing (e.g, $\omega_{21} < \cdots < \omega_{2J}$). The reason that it is reasonable to assume non-informative prior $\omega_{0j} = \omega_{1j} = \omega_{2j} = 1/3$ hinges on the fundamental characteristic of dose finding: it is a sequential decision-making process of testing the dose in the *one-at-a-time fashion from low to high doses*. We escalate to the next higher dose only when interim data show that the current dose is safe. For instance, when data (e.g., 0/3 DLT) show that d_1 is safe, we are ready to escalate to d_2. At that moment, although we know d_2 is more toxic than d_1, we do not know by how much. Compared to d_1, *a priori*, it is not clear that d_2 is just slightly more toxic and still deemed safe, or more toxic and close to the MTD, or substantially more toxic and thus above the MTD. In other words, during the trial, for each new dose to be escalated to for testing, *a priori*, we do not know if it is underdosing, proper dosing, or overdosing. Thus, the most sensible approach is to assume that d_j, $j = 1, \cdots, J$, has an equal probability of being underdosing, proper dosing and overdosing (i.e., $\omega_{0j} = \omega_{1j} = \omega_{2j} = 1/3$). Actually, the dose escalation rule itself (i.e., escalate when the current dose is safe) already (implicitly) accounts for the prior information that a higher dose is expected to be more toxic. There is no need to do it again by imposing an informative prior. Liu and Yuan (2015) evaluated the use of an informative prior (e.g., $\omega_{21} < \cdots < \omega_{2J}$) and found that overall it does not improve performance. Of note, when the trial is completed, to evaluate the dose-toxicity relationship, BOIN does account for the monotonicity to estimate π_j by isotonic regression.

3. BOIN makes dose escalation/de-escalation decisions based only on the local data at the current dose. Does this seemly "myoptic" approach cause any notable loss of efficiency?

The answer is no. For the purpose of dose finding, the loss of efficiency due to the use of local data is minimal, and mostly ignorable. This is because unlike most statistical inferential procedures, dose finding is a sequential decision-making process, escalating from low doses to high doses. Suppose that the current dose level is j. In order to reach j, the data observed previously at lower doses (i.e., $< j$) must indicate that these doses are safe and substantially lower than the MTD (e.g., 0/3 DLT). Thus, these data provide little information to determine whether the current dose j is below, equal (or sufficiently close) to, or above the MTD to make the decision of dose escalation/de-escalation. Similarly, in order to reach j, the data observed previously at higher doses (i.e., $> j$), if exist, must indicate that these doses are substantially higher than the MTD (e.g., 2/3 DLT). Thus, borrowing information from these data helps little to determine if the current dose j is below, equal (or sufficiently close) to, or above the MTD. In other words, the information used to evaluate the safety of dose j and make the decision of dose escalation/de-escalation is mostly provided by the local data observed at dose j. This explains why in general BOIN yields comparable performance as CRM that uses a dose-toxicity model to borrow information across doses, as demonstrated in the next section. Lin and Yuan (2019) showed that incorporating the data other than the current dose into BOIN provides virtually no efficacy gain.

Although BOIN uses local data to make the decisions of dose escalation/de-escalation, it indeed uses all data to select the MTD based on nonparametric isotonic regression. As noted by Liu and Yuan (2015), dose transition and MTD selection actually are two independent components of a dose finding study. When the trial is completed, any method (e.g., a logistic model), can also be used with BOIN to select the MTD. The parametric model yields better MTD selection when the model is correctly specified, but may perform poorly when the model is misspecified. Overall, isotonic regression provides a robust and competitive approach.

4. What is the relationship between the keyboard/mTPI-2 design and BOIN? Which one is better?

The keyboard/mTPI-2 design can be regarded as a special convoluted version of BOIN. Keyboard/mTPI-2 makes decisions of dose escalation/de-escalation based on the location of the strongest key (i.e., the interval with the largest posterior probability), with respect to the target key/interval (δ_1, δ_2). As each key is of equal length and the beta distribution is unimodal, this is virtually equivalent to pinpoint where the mode of $f(\pi_j|D_j)$ is, with respect to (δ_1, δ_2). It is not exactly equivalent because the beta distribution is asymmetric. According to equation (3.2), when the unform prior $\pi_j \sim \text{Beta}(1,1)$ is assumed, the mode of $f(\pi_j|D_j)$ is simply $\hat{\pi}_j = y_j/n_j$. As a result, the dose escalation/de-escalation/stay decision of keyboard/mTPI-2 boils down to whether $\hat{\pi}_j \leq \delta_1$, $\hat{\pi}_j > \delta_2$, or $\hat{\pi}_j \in (\delta_1, \delta_2]$, which essentially is the BOIN decision rule. This intrinsic link explains why keyboard/mTPI-2 often has similar performance as BOIN (see the next section for a numerical study).

There are, however, several important differences between keyboard/mTPI-2 and BOIN. First, BOIN is more transparent and straightforward, as described previously. Second, (δ_1, δ_2) in keyboard/mTPI-2 (when the uniform prior Beta(1, 1) is assumed) plays the same role as dose escalation/de-escalation boundaries (λ_e, λ_d) in BOIN. However, the latter is optimized to minimize the decision error, while the former is subjectively specified and lack of clear statistical properties and assurance. Third, the operating characteristics of keyboard/mTPI-2 critically rely on the prior assumption that $\pi_j \sim \text{Beta}(1,1)$. Under the beta-binomial model, this prior assumes a prior sample size of two patients and the prior DLT probability of 0.5 for each dose, which may be overly informative (given that some doses often only treat 3 patients) and unrealistically toxic. Often, one may prefer a vague prior such as Beta(0.02, 0.08), which is equivalent to a prior sample size of 0.1 patient with a prior DLT probability of 0.2, to allow observed data to drive decisions. Unfortunately, when this more reasonable vague prior is used, the operating characteristics of keyboard/mTPI-2 become elusive (e.g., resulting in overly aggressive dose escalation) (Zhou et al., 2021a). In this case, the mode of $f(\pi_j|D_j)$ is not $\hat{\pi}_j$ anymore, and the decision rule of keyboard/mTPI-2 diverges from the BOIN decision rule. In contrast, BOIN does not require the prior assumption $\pi_j \sim \text{Beta}(1,1)$. It only assumes that for any new dose to be tested, it has an equal probability of being below, equal (or sufficiently close) to, or above the MTD, see the answer to question 3 for why it is a reasonable prior assumption of equipoise. BOIN's dose elimination rule employs a beta-binomial model with the uniform prior $\pi \sim \text{Beta}(1,1)$, see Section 3.4.1. However, for the purpose of dose elimination (or overdose control), using Beta(1, 1) prior is not critical. When another prior is used (e.g., Beta(0.02, 0.08)), the overdose control probability cutoff 0.95 can be slightly adjusted (e.g., to 0.93) to obtain same operating characteristics. Therefore, BOIN is a preferred, simpler, and more robust choice of design.

3.5 Operating characteristics

Zhou et al. (2018b) conducted comprehensive Monto Carlo studies to compare the operating characteristics of various phase I designs, including the 3+3 design, three model-based designs (CRM, EWOC, and BLRM), and three model-assisted designs (mTPI, keyboard design, and BOIN). In addition, variations of the CRM and BLRM designs, including CRM-DS (allowing dose skipping) and BLRM-NOC (without overdose control), were also investigated. The target DLT probability is $\phi = 0.25$, with $J = 6$ dose levels and a maximum sample size of $N = 36$. The starting dose level is $j = 1$. Patients are treated in cohorts of size three. The proper dosing interval is $(\delta_1, \delta_2) = (\phi - 0.05, \phi + 0.05)$ for the mTPI, keyboard, and BLRM designs, and the BOIN design uses $\phi_1 = 0.6\phi$ and $\phi_2 = 1.4\phi$ as recommended by these designs. The 3+3 design often completes before reaching the prespecified maximum sample size (e.g., when 2/3 or 2/6 had DLT). For fair comparisons, after the 3+3 design selects MTD, an expansion cohort is treated at MTD to reach a total sample size of 36. The detailed configurations for these designs and more simulation results (e.g., $\phi = 0.2$ or 0.3, or cohort size $= 1$) can be found in Zhou et al. (2018b).

To avoid cherry-picking and inadvertent selection biases, 1000 true dose–toxicity scenarios (or curves) were randomly generated using the pseudo-uniform algorithm (Clertant and O'Quigley, 2017) for comparison. Given a

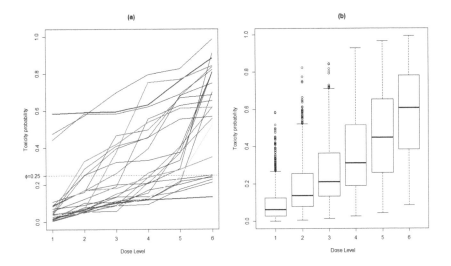

FIGURE 3.4: Panel (a): 25 randomly selected dose–toxicity curves with six picked curves showing different shapes; Panel (b): Distribution of the DLT probabilities by dose level from the 1000 randomly generated scenarios.

target DLT rate ϕ and J dose levels, the random scenarios were generated as follows:

1. Select one of the J dose levels as the MTD with equal probabilities.

2. Sample $M \sim \text{Beta}(\max\{J - j, 0.5\}, 1)$, where j denotes the selected MTD level, and set an upper bound $B = \phi + (1 - \phi)M$ for the toxicity probabilities.

3. Repeatedly sample J toxicity probabilities uniformly on $[0, B]$ until these correspond to a scenario in which dose level j is MTD.

Figure 3.4 panels (a) and (b) show 25 randomly selected scenarios and distributions of the DLT probabilities by dose level from the 1000 scenarios, respectively. It can be seen that the simulated dose–toxicity curves cover various shapes and a wide range of toxicity probabilities. The algorithm above guarantees that the generated dose–toxicity curves are monotonically increasing (i.e., higher doses have higher toxicity rates). For each scenario, 2000 trials were simulated.

Performance metrics

- *Accuracy*
  A1. The percentage of correct selection (PCS), defined as the percentage of simulated trials in which the target dose is correctly selected as MTD. When

all the dose levels are above MTD (i.e., the DLT probability of the lowest dose $> \phi + 0.1$), PCS is defined as the percentage of early termination of trials.

A2. The average percentage of patients who are assigned to MTD across the simulated trials. When all the dose levels are above MTD (i.e., the DLT probability of the lowest dose $> \phi + 0.1$), the average percentage of patients not enrolled into the trial is used for this metric.

- *Safety*

  B1. The percentage of simulated trials in which a toxic dose with the true DLT probability $> 33\%$ is selected as MTD when the target $\phi = 25\%$.

  B2. The average percentage of patients assigned to the toxic doses with true DLT probabilities $> 33\%$ when the target $\phi = 25\%$.

- *Reliability*

  C1. The risk of overdosing, defined as the percentage of simulated trials with more than 50% of patients treated at doses above MTD.

  C2. The risk of poor allocation, defined as the percentage of simulated trials in which fewer than six patients are treated at MTD.

  C3. The risk of irrational dose assignment, defined the percentage of times that the design fails to de-escalate the dose when 2/3 or $> 3/6$ patients had DLTs at a dose.

Reliability metrics C1 to C3 measure the likelihood of a design demonstrating problematic behaviors (e.g., treating 50% or more patients at toxic doses, or fewer than six patients at MTD) that have severe clinical consequences. These metrics are of great practical importance, but unfortunately are often overlooked in the literature. The reliability metrics are not covered by other metrics. For example, the percentage of patients overdosed (i.e., metric B2) does not cover the risk of overdosing (i.e., metric C1). Two designs can have similar percentages of patients overdosed, but rather different risks of overdosing 50% of the patients. Statistically, metric B2 measures the mean of overdosing, while metric C1 measures the tail probability of overdosing.

Results

Accuracy   Panels A1 and A2 in Figure 3.5 show distributions of the PCS and the average percentages of patients treated at MTD, respectively, for the investigational designs relative to the 3+3 design across 1000 scenarios. That is, the values displayed in the figure are the difference between those of a specified design and the reference (i.e., 3+3 design). For example, PCS = 0 means that the design has the same PCS as the 3+3 design. As each dose–toxicity scenario generates a value of the performance metric (e.g., PCS), there are a total of 1000 values for each of the metrics across the 1000 scenarios. The boxplot reflects the distribution of the metric across the 1000 scenarios. In terms of the accuracy of correctly selecting the MTD, the CRM, mTPI, BOIN, and keyboard designs are comparable and substantially outperform the 3+3 design. The BLRM and EWOC designs perform the worst, with the average PCS similar to that of the 3+3 design. The EWOC design also has the largest variation in PCS. The results for the number of patients treated at MTD are similar to those for PCS. The CRM, mTPI, BOIN, and keyboard designs are generally comparable and substantially outperform the 3+3 design. The mTPI and CRM designs allocate slightly more patients to MTD than the BOIN and keyboard designs, but the latter two designs are less variable, as shown by the shorter boxes in the box plot. mTPI is less robust than the BOIN and keyboard designs. For example, when the target $\phi = 20\%$, mTPI has notably lower PCS than the BOIN and keyboard designs, see Zhou et al. (2018b) for details.

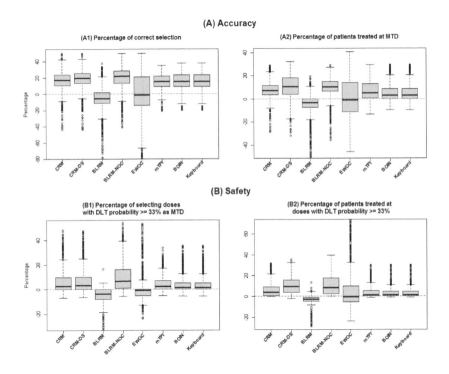

FIGURE 3.5: Accuracy and safety of the eight designs with respect to the 3+3 design, including (A1) percentage of correct selection of MTD, (A2) percentage of patients treated at MTD, (B1) percentage of selecting a dose with the DLT probability $\geq 33\%$ as MTD, and (B2) percentage of patients treated at doses with DLT probabilities $\geq 33\%$. For (A1) and (A2), a larger value indicates better performance; a positive value means that the design outperforms the 3+3 design. For (B1) and (B2), a smaller value indicates better performance; a negative value means that the design outperforms the 3+3 design.

Safety   As shown in Figure 3.5 panel B1, the CRM, mTPI, BOIN, and keyboard designs are comparable in terms of the percentage of selecting a toxic dose (with a DLT probability $\geq 33\%$) as MTD, but CRM and mTPI are slightly more variable than the BOIN and keyboard designs. The BLRM and EWOC designs are the most conservative and least likely to select a toxic dose as MTD. In terms of the percentage of patients treated at a toxic dose with a DLT probability $\geq 33\%$, on average the CRM, mTPI, BOIN, and keyboard designs are comparable, but BOIN and keyboard show smaller variations (Figure 3.5 panel B2).

Reliability   In terms of the risk of overdosing 50% or more of the patients (Figure 3.6, panel C1), the BLRM, BOIN, and keyboard designs perform the best. The performances of the CRM and mTPI designs are similar and rank

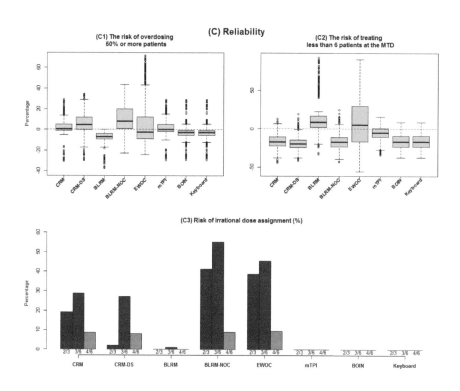

FIGURE 3.6: Reliability of the eight designs with respect to the 3+3 design, including (C1) risk of overdosing 50% or more patients, (C2) risk of treating < 6 patients at the MTD, and (C3) risk of irrational dose assignments. A smaller value indicates better performance; negative value means that the design outperforms the 3+3 design.

in between the performances of these other designs. The EWOC design has a similar averaged risk of overdosing patients as the BOIN and keyboard designs, but is much more variable. Of note, the CRM, mTPI, BOIN, and keyboard designs, on average, overdose similar percentages of patients (Figure 3.6 panel B2), but have different risks of overdosing 50% or more of the patients (Figure 3.6, panel C1). This indicates that the risk of overdosing (50% or more patients) and the average percentage of patients overdosed indeed measure different aspects of a design, and it is thus important to consider both metrics when evaluating a design. In terms of the risk of poor allocation (i.e., treating fewer than six patients at the MTD, see Figure 3.6, panel C2), BLRM and EWOC perform the worst, with a significantly higher risk than the other designs. The CRM, BOIN, and keyboard designs have comparable risks of poor allocation.

In terms of the risk of irrational dose assignment (Figure 3.6, panel C3), the model-assisted designs outperform the model-based designs. The model-based designs (i.e., CRM, BLRM, and EWOC) have an 8% to 55% chance of failing to de-escalate the dose when $\geq 2/3$ or $\geq 3/6$ patients have DLTs, whereas such irrational dose assignments never occur in the mTPI, BOIN, and keyboard designs. The model-based designs rely on the assumed model to make the decision of dose assignment. When the model is misspecified, the estimates can be biased and thus irrational dose assignment arises. The model-assisted designs are free of that issue because they do not impose any model assumption on the dose–toxicity curve. For example, by its dose escalation/de-escalation rule, the BOIN design guarantees de-escalating the dose if the observed DLT rate at the current dose is higher than 29.8%, given the target DLT rate of 25%.

In summary, the model-assisted designs (e.g., the BOIN and keyboard designs) substantially outperform the algorithm-based 3+3 design in the accuracy of identifying MTD and allocating patients to MTD. They produce competitive accuracy and safety comparable to the model-based designs (e.g., CRM), but are much simpler and more transparent. In addition, the model-assisted designs are more robust, and avoid the irrational dose assignment of the model-based designs due to model misspecification. Among the model-assisted designs, BOIN stands out. It has similar operating characteristics as the keyboard design, but is simpler, more flexible, and transparent. The mTPI design is not recommended due to the poor reliability and safety concerns.

3.6 Software and case study

Software
Software for the BOIN design is available in several different forms: (1) Web application "BOIN Suite" with intuitive graphical user interface, freely

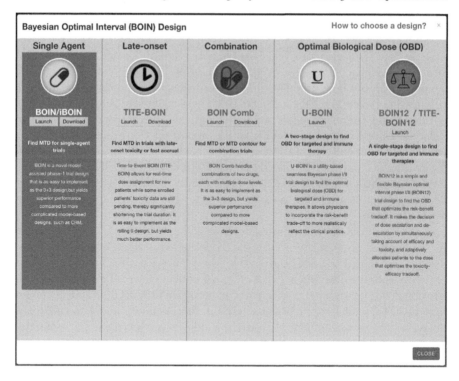

FIGURE 3.7: The launchpad of the BOIN Suite, freely available at `http://www.trialdesign.org`.

available at `http://www.trialdesign.org`. Figure 3.7 shows the launchpad of BOIN Suite, which is a software platform to handle various types of early-phase clinical trials based on the unified family of the BOIN design and its extensions. Figure 3.8 shows the decision tree to assist users to choose an appropriate BOIN design module for different types of trials; (2) R package `BOIN` freely available at `CRAN`; and (3) Windows desktop program BOIN Suite, freely available at the MD Anderson Cancer Center Software Download Website `https://biostatistics.mdanderson.org/softwaredownload/`. We recommend the web application as it is most frequently updated.

The keyboard design can be implemented using the web application "Keyboard Suite," freely available at `http://www.trialdesign.org`. The R script for implementing the mTPI design can be downloaded from the MD Anderson Cancer Center Software Download Website `https://biostatistics.mdanderson.org/softwaredownload/`.

Case study
Solid Tumor Dose Finding Trial   The objective of this phase I trial (ClinicalTrials.gov Identifier: NCT03725436) was to determine MTD for the

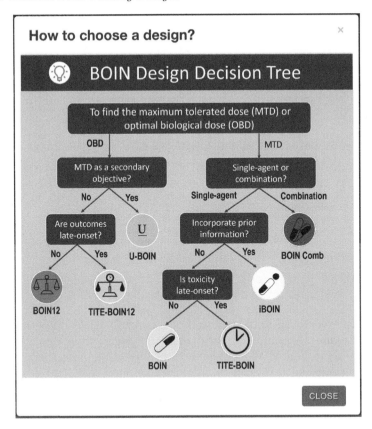

FIGURE 3.8: The decision tree of the BOIN Suite software to assist users to choose an appropriate BOIN design module.

MDM2/MDMX inhibitor ALRN-6924 in combination with paclitaxel in adult patients with advanced or metastatic solid tumors with wild-type (WT) TP53. The trial employed the BOIN design to find MTD. Five doses of ALRN-6924 were investigated with the target DLT probability $\phi = 25\%$.

We use this trial as an example to illustrate the use of the BOIN software, and provide guidance to address some common design issues in phase I trials. The Keyboard Suite web application shares a similar user interface as the BOIN Suite, thus we here focus on the latter. After selecting and launching the "BOIN/iBOIN" module from the BOIN Suite launchpad, we design the trial using the following three steps:

Step 1: Enter trial parameters
Doses and Sample Size   As shown in Figure 3.9, the number of doses under investigation is five; the starting dose level is one. For trials where the lowest

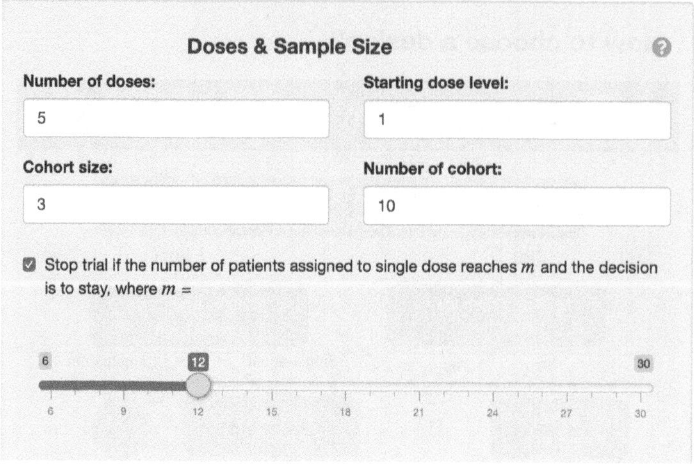

FIGURE 3.9: Specify doses, sample size, and convergence stopping rule.

dose is believed to be safe, starting from a slightly higher dose level (e.g., two) may reduce the sample size as it allows the design to reach MTD sooner. However, due to limited knowledge on the safety of the new drug, in general it is not recommended to start from a high dose level (e.g., four).

The cohort size for the trial is three and the number of cohorts is 10, with a total sample size of 30. As a rule of thumb, we recommend the maximum sample size $N = 6 \times J$ (i.e., the maximum sample size of the 3+3 design) as the total sample size, where J is the number of doses. This sample size generally yields reasonable operating characteristics (e.g., 50–70% correct selection percentage of the true MTD). To reduce the sample size, it is often useful to use the "convergence" stopping rule: stop the trial early when m patients have been assigned to a dose and the decision is to stay at that dose. This stopping criterion suggests that the dose finding approximately converges to MTD, thus the trial can be stopped. We recommend $m = 9$ or larger. In this trial, $m = 12$ is used. Because of the early stopping rule, the actual sample size used in the trial is often smaller than the prespecified maximum sample size N. The saving depends on the true dose–toxicity scenario and can be evaluated using simulation. Usually, the savings in the sample size is more prominent when the true MTD is near the starting dose.

The choice of the cohort size should be based on considerations of design performance and logistic complexity. Given a fixed total sample size, the use of large cohort sizes will result in fewer cohorts. This reduces the logistical burden and requires fewer dose escalation/de-escalation decisions for trial conduct. As a tradeoff, using a large cohort size often reduces the accuracy of identifying the MTD and evaluating trial safety, because trials based on a smaller number of cohorts tend to be less adaptive. For example, given a total sample size of 30 patients, using a cohort size of six patients results in a total of five cohorts. This means that during the trial, we only have four chances to escalate/de-escalate the dose. If MTD is dose level five or six, and if the dose escalation starts from dose level one and no dose skipping is permitted, then there is a high likelihood that we will not be able to reach MTD because many patients are treated in lower doses. In addition, using a cohort size of six may expose up to six patients to an overly toxic dose before dose de-escalation is made. In contrast, using small cohort sizes, such as one or two patients per cohort, renders the trial more freedom to move up and down between doses and be more responsive/adaptive to the observed data. This is logistically more complicated, however, because both data and dose escalation/de-escalation decisions need to be updated more frequently. In addition, it prolongs the trial duration as new patients cannot be enrolled until the patient in the previous cohort has completed their DLT assessment. As a result, the most commonly-used cohort size in practice is two to four patients per cohort.

Target and Accelerated Titration   As shown in Figure 3.10, the target DLT probability is $\phi = 0.25$, which should be elicited from clinicians. The target ϕ can be adjusted if a specific safety requirement is desirable. For example, if it is desirable to de-escalate the dose when the DLT rate is > 30%, then $\phi = 25\%$ is an appropriate target with the de-escalation boundary $\lambda_d = 0.298$. As described previously, such simple mapping between the target and escalation/de-escalation boundaries is a unique and important advantage of BOIN.

Given ϕ, although the software allows users to specify ϕ_1 and ϕ_2 (see Sections 3.4.2 and 3.4.3 for their definition and interpretation) by unchecking the

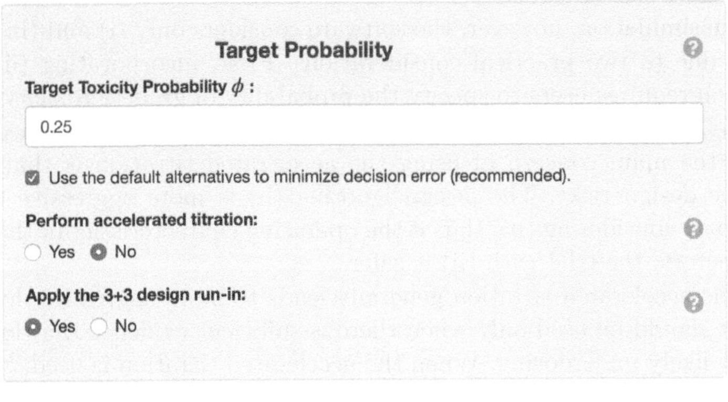

FIGURE 3.10: Specify the target, accelerated titration, and 3+3 design run-in (available only when the target $\phi = 0.25$).

box "☑ use the default alternatives to minimize decision error" under the target toxicity probability field, we highly recommend using the default values (i.e., $\phi_1 = 0.6\phi$ and $\phi_2 = 1.4\phi$) provided by the software. These default values have been shown to produce highly robust and desirable operating characteristics. Nevertheless, when necessary, the values of ϕ_1 and ϕ_2 can be calibrated to satisfy certain trial design goals. For example, if we do not want to modify ϕ and prefer a more conservative design, we may set $\phi_2 = 1.2\phi$ to obtain a lower de-escalation boundary λ_d. To this end, it is important to distinguish (ϕ_1, ϕ_2) from (λ_e, λ_d), where ϕ_1 and ϕ_2 are the toxicity probabilities used to minimize the decision errors, and λ_e and λ_d are the decision boundaries actually used to determine dose escalation and de-escalation. A practical way to judge if (ϕ_1, ϕ_2) specified by users is appropriate or not is to examine whether the resulting (λ_e, λ_d) makes practical and clinical sense. For example, given $\phi = 0.25$, if we set $(\phi_1, \phi_2) = (0.9\phi, 1.1\phi)$, the resulting $(\lambda_e, \lambda_d) = (0.237, 0.262)$ makes little sense because λ_e and λ_d are too close and nearly indistinguishable under small sample sizes. The fundamental issue here is that it is not meaningful to minimize the decision error for the hypotheses (i.e., ϕ vs. $\phi_1 = 0.9\phi$ vs. $\phi_2 = 1.1\phi$) given that we have no power to distinguish. See Section 3.4.3 for more discussion.

In addition to the "convergence" stopping rule described above, another useful approach offered by the BOIN software to reducing the sample size is to conduct the accelerated titration (Simon et al., 1997) before treating patients in cohorts of three (Figure 3.10). During the accelerated titration, we treat patients in cohorts of one, and we continue escalating the dose in the one-patient-per-dose-level fashion until any of the following events are observed: (i) the first instance of DLT, (ii) the second instance of moderate (grade 2) toxicity, or (iii) the highest dose level is reached. At that point, the titration ends. We add two more patients to the current dose level, and hereafter switch to the cohort size of three.

In the simulation, however, the software considers only (i) and (iii), ignoring (ii), due to two practical considerations. First, incorporating (ii) in the simulation requires users to specify the probability of grade 2 toxicity at each dose level, which is cumbersome and brings substantial noise and uncertainty. Second, the main concern of using the accelerated titration is that it may make the design risky. The design ignoring (ii) is more aggressive than its counterpart considering (ii), thus if the operating characteristics of the former is satisfactory, the trial with (ii) is safer.

As the accelerated titration generally leads to more aggressive dose escalation, it should be used only when there is sufficient evidence that low doses are most likely underdosing. When the accelerated titration is used, based on our experience, a reasonable rule of thumb for the maximum sample size is $N = j^* - 1 + 6(J - j^* + 1)$, where j^* is the dose level where the first DLT is expected to occur. For example, if we expect that the first three doses are very safe and the first DLT may occur in dose level $j^* = 4$, then the maximum required sample size N could be reduced to 15 when there are five dose levels.

When the target $\phi = 0.25$, the software provides an option "□Apply the 3+3 design run-in" to embed the 3+3 design rule into the BOIN design. The rationale for this option is that when $\phi = 0.25$, the default BOIN de-escalation boundary is $\lambda_d = 0.298$, which means de-escalating the dose when 1/3 or 2/6 patients have DLT. Due to the influence of the conventional 3+3 design, in some cases, investigators prefer that the design stays at the current dose when 1/3 patients have DLT. The 3+3 run-in option enables that and enforces the design to stay at the current dose when 1/3 DLT. This option is added mainly based on practical consideration, not statistical consideration. Actually, the 3+3 design rule—stay when 1/3 DLT, but de-escalate when 2/6 DLT—is not self-consistent, see Section 3.4.4 for the explanation of why using the same fixed de-escalation boundary is optimal. Nevertheless, as the modification only occurs when the cumulative number of patients is three, activating the option generally has minor impact on the design operating characteristics.

Overdose Control   This panel (see Figure 3.11) specifies the overdose control rule, described in Section 3.4.1. That is, if $\Pr(\pi_j > \phi \mid D_j) > 0.95$ and $n_j \geq 3$, dose level j and higher are eliminated from the trial, and the trial is terminated if the lowest dose level is eliminated. When the trial is terminated due to toxicity, no dose should be selected as MTD. In general, we recommend using the default probability cutoff 0.95. A smaller value (e.g., 0.9), results in stronger overdose control, but at the cost of reducing the probability of correctly identifying MTD. This is because, in order to correctly identify MTD, it is imperative to explore the doses sufficiently to learn their toxicity profile.

In some trial settings, under the null case that all doses are overly toxic (i.e., the lowest/first dose is above the MTD), the probability of early trial termination may not be as high as we desire (e.g., > 70%). That is simply

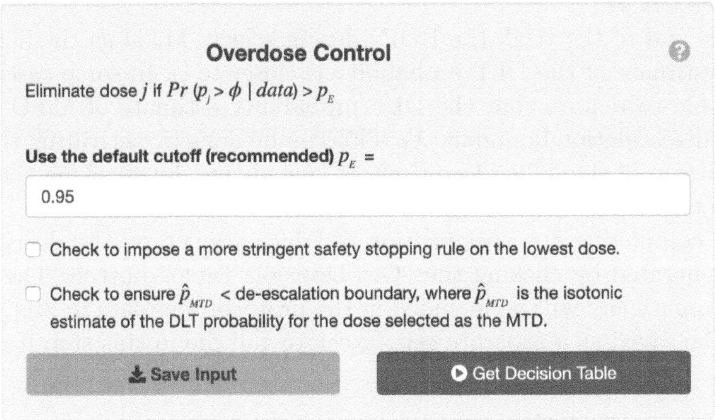

Overdose Control   ❓

Eliminate dose j if $Pr\,(p_j > \phi \mid data) > p_E$

Use the default cutoff (recommended) p_E =

0.95

☐ Check to impose a more stringent safety stopping rule on the lowest dose.

☐ Check to ensure $\hat{p}_{MTD}$ < de-escalation boundary, where $\hat{p}_{MTD}$ is the isotonic estimate of the DLT probability for the dose selected as the MTD.

⬇ Save Input	▶ Get Decision Table

FIGURE 3.11: Specify the overdose control rules.

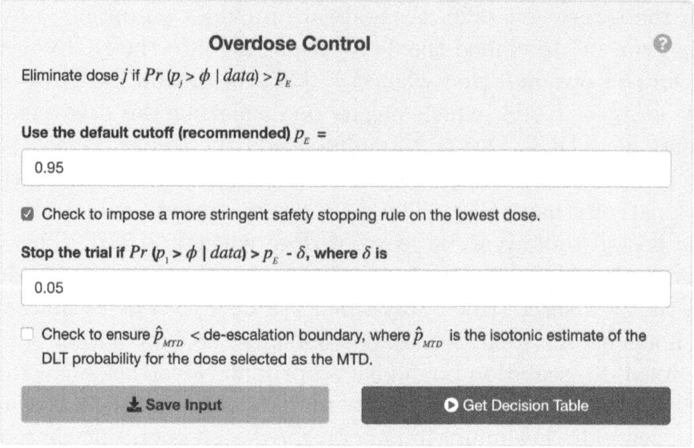

FIGURE 3.12: Adjust the overdose control cutoff.

because the small sample size cannot provide enough power to distinguish whether a dose is overly toxic or not (e.g., 40% vs. 30%). To achieve a high early termination probability (when the first dose is overly toxic), we can activate the first option "□ Check to impose a more stringent safety stopping rule on the lowest dose" (see Figure 3.12). This option makes the lowest/first dose more likely to be eliminated by lowering the probability cutoff by δ. The default value $\delta = 0.05$ is generally recommended and produces a good balance between the safety and the accuracy to identify MTD. A large value of δ (e.g., 0.1) increases the early termination probability (when all doses are toxic), but at the cost of reducing the probability of correctly identifying MTD when it is the lowest dose.

At the end of the trial, the BOIN design selects MTD as the dose whose isotonic estimate of the DLT probability is closet to ϕ. In some cases, it may be desirable to require that the DLT probability estimate of MTD be lower than the de-escalation boundary λ_d. This can be done by activating the option at the bottom of the "Overdose Control" panel (i.e., by checking the box "□ Check to ensure $\hat{p}_{MTD} <$ de-escalation boundary.")

After completing the specification of trial parameters, the decision table will be generated by clicking the "Get Decision Table" button. The decision table automatically will be included in the protocol template in Step 3, but it can also be saved as a separate csv, Excel, or pdf file in this step if needed.

Step 2: Run simulation

Operating Characteristics   This step generates the operating characteristics of the design through simulation, see Figure 3.13. The scenarios used for simulation should cover various possible clinical scenarios (e.g., MTD is located

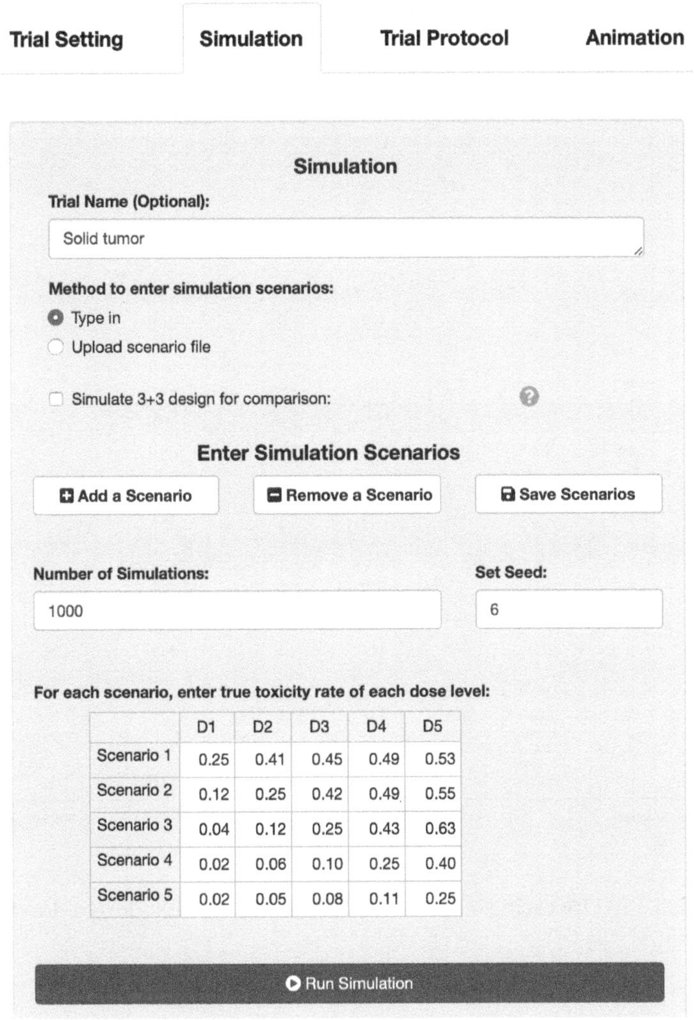

Trial Setting   **Simulation**   **Trial Protocol**   **Animation**

Simulation

Trial Name (Optional):

Solid tumor

Method to enter simulation scenarios:

○ Type in
○ Upload scenario file

☐ Simulate 3+3 design for comparison: ❓

Enter Simulation Scenarios

| ➕ Add a Scenario | ➖ Remove a Scenario | 💾 Save Scenarios |

Number of Simulations: **Set Seed:**

| 1000 | 6 |

For each scenario, enter true toxicity rate of each dose level:

	D1	D2	D3	D4	D5
Scenario 1	0.25	0.41	0.45	0.49	0.53
Scenario 2	0.12	0.25	0.42	0.49	0.55
Scenario 3	0.04	0.12	0.25	0.43	0.63
Scenario 4	0.02	0.06	0.10	0.25	0.40
Scenario 5	0.02	0.05	0.08	0.11	0.25

▶ Run Simulation

FIGURE 3.13: Simulate operating characteristics of the design.

at different dose levels). To facilitate the generation of the operating characteristics of the design, the software automatically provides a set of randomly generated scenarios with various MTD locations, which are often adequate for most trials. Depending on the application, users can add or remove scenarios. The software also provides an option to include the 3+3 design as a comparator to facilitate the communication with clinicians who are more familiar with the conventional design. Figure 3.14 shows the operating characteristics of BOIN for the solid tumor trial. The simulation results will be automatically

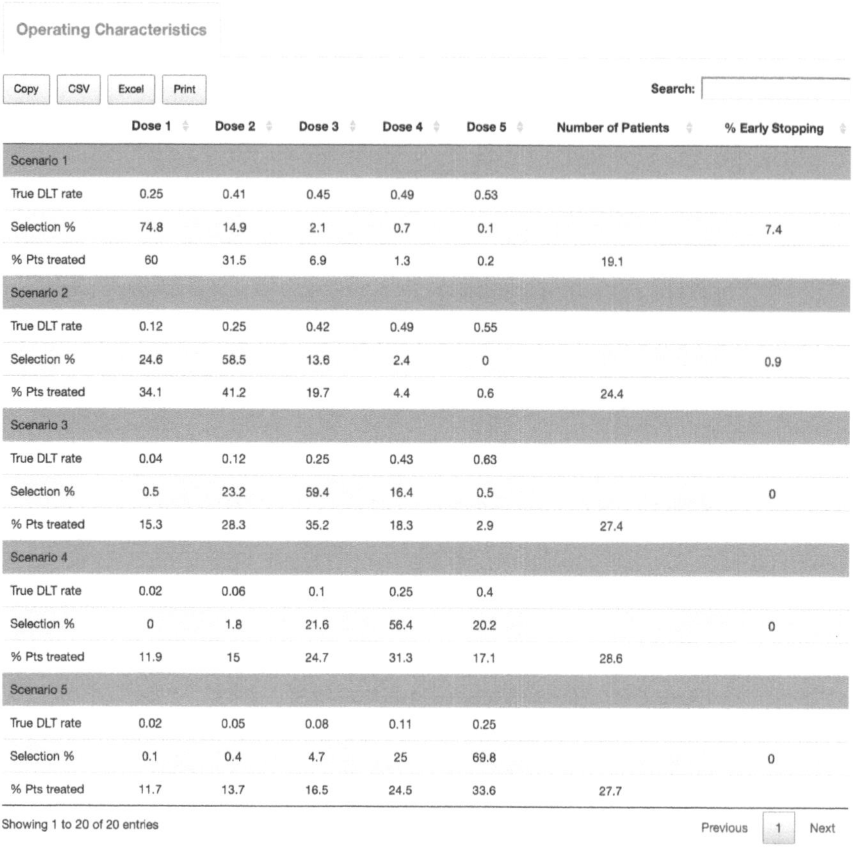

Operating Characteristics

| | Copy | CSV | Excel | Print | | | Search: |

	Dose 1	Dose 2	Dose 3	Dose 4	Dose 5	Number of Patients	% Early Stopping
Scenario 1							
True DLT rate	0.25	0.41	0.45	0.49	0.53		
Selection %	74.8	14.9	2.1	0.7	0.1		7.4
% Pts treated	60	31.5	6.9	1.3	0.2	19.1	
Scenario 2							
True DLT rate	0.12	0.25	0.42	0.49	0.55		
Selection %	24.6	58.5	13.6	2.4	0		0.9
% Pts treated	34.1	41.2	19.7	4.4	0.6	24.4	
Scenario 3							
True DLT rate	0.04	0.12	0.25	0.43	0.63		
Selection %	0.5	23.2	59.4	16.4	0.5		0
% Pts treated	15.3	28.3	35.2	18.3	2.9	27.4	
Scenario 4							
True DLT rate	0.02	0.06	0.1	0.25	0.4		
Selection %	0	1.8	21.6	56.4	20.2		0
% Pts treated	11.9	15	24.7	31.3	17.1	28.6	
Scenario 5							
True DLT rate	0.02	0.05	0.08	0.11	0.25		
Selection %	0.1	0.4	4.7	25	69.8		0
% Pts treated	11.7	13.7	16.5	24.5	33.6	27.7	

Showing 1 to 20 of 20 entries   Previous | 1 | Next

FIGURE 3.14: Operating characteristics of the BOIN design for the solid tumor trial.

included as a table in the protocol template in the next step, but can also be saved as a separate csv or Excel file if needed.

Step 3: Generate protocol template
Protocol Preparation   The BOIN software generates sample texts and a protocol template (including the simulation results in Step 2) to facilitate the protocol write-up. The protocol template can be downloaded in various formats (see Figure 3.15). Use of this module requires the completion of Steps 1 and 2. Once the protocol is approved by regulatory bodies (e.g., Institutional Review Board), we follow the design decision table to conduct the trial and

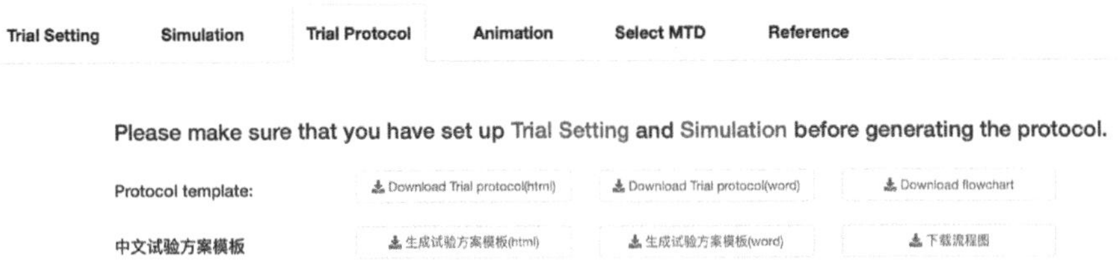

FIGURE 3.15: Download protocol templates.

make adaptive decisions (e.g., dose escalation/stay/de-escalation). When the trial completes, use the "Select MTD" module to select MTD. The software outputs the recommended MTD, the posterior estimate of DLT probability at each dose, and a 95% credible interval (see Figure 3.16).

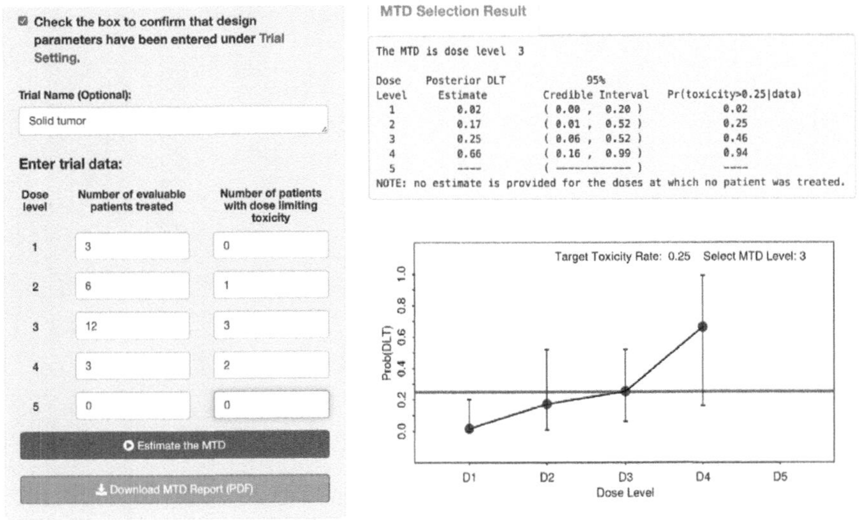

FIGURE 3.16: Estimate and identify MTD at the completion of the solid tumor trial based on a hypothetical dataset.

4

Drug-Combination Trials

4.1 Introduction

Drug combination therapy provides an effective approach to improving treatment efficacy and overcoming most cancers' resistance to monotherapy. The objectives of using drug combinations are to induce an additive or a synergistic treatment effect, increase the joint dose intensity with non-overlapping toxicities, and target various tumor cell susceptibilities and disease pathways. Despite the enormous importance of combination therapies, statistical designs currently used for dose finding in phase I trials of combination therapies are grossly inefficient and rudimentary—most combination trials have used the conventional 3+3 design (Riviere et al., 2015a). The objective of this chapter is to address the challenges and clarify misconceptions in designing combination therapy trials and introduce more efficient designs, in particular model-assisted designs, for dose finding in phase I drug-combination trials.

In general, drug combinations may involve one of the followings: two or more previously marketed drugs or biologics, two or more new molecular entities, or a mix of previously marketed drugs or biologics and new molecular entities. According to the US Food and Drug Administration (FDA) guidelines (FDA, 2006, 2013), prior to testing a new drug combination in human beings, extensive preclinical studies are required to demonstrate the biological rationale for the combination and to assess the safety of the combination (FDA, 2006). When such data are not available or indicate safety concerns for the combination, additional toxicology studies are required to address the concerns. Sometimes drug-combination trials may involve two or more new investigational drugs that have not been previously studied for any indication. In such cases, additional considerations are needed for the co-development of the new investigational drugs for use in combination (FDA, 2013). In drug-combination trials, it is useful to test multiple doses of each drug to identify the optimal dose combination in terms of risks and benefits (FDA, 2013). Compared to single-agent trials, drug-combination trials have a higher dimension for the dose searching space, leading to several unique challenges.

The major challenge in designing combination trials is that the dose combinations under investigation are only partially ordered by the dose-limiting toxicity (DLT) probability). This is in contrast with monotherapy trials, for which the doses under investigation are fully ordered by the DLT probability

DOI: 10.1201/9780429052781-4

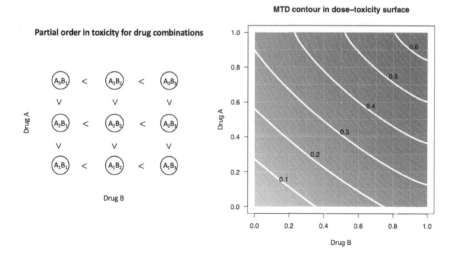

FIGURE 4.1: Partial ordering (left) and toxicity contours (right) for drug combinations.

(i.e., the higher the dose, the greater the DLT probability). Consider a trial combining J doses of agent A, denoted as $A_1 < A_2 < \cdots < A_J$, and K doses of agent B, denoted as $B_1 < B_2 < \cdots < B_K$. Let $A_j B_k$ denote the combination of A_j and B_k. Often it is reasonable to assume that when the dose of agent A is held constant, the DLT probability for the combination increases in the dose of agent B, and vice versa. As shown in the left panel of Figure 4.1, the rows and columns of the dose combination matrix are partially ordered, with the DLT probability increasing in the dose of the corresponding agent when the dose of the other agent is fixed. However, in other directions of the dose matrix (e.g., along the diagonals from the upper left corner to the lower right corner), the ordering is not clear due to unknown drug-drug interactions. For example, *a priori* we do not know whether the DLT probability for $A_2 B_2$ is higher than the DLT probability for $A_1 B_3$ or $A_3 B_1$.

The partial ordering of dose combinations has several implications. First, monotherapy designs for finding the maximum tolerated dose (MTD), described in previous chapters, cannot directly be used for finding MTD of a drug combination. The second important implication is that there is not just a single MTD. Rather, as depicted in the right panel of Figure 4.1, there is an MTD contour in the two-dimensional dose space. Therefore, multiple MTDs may exist in the $J \times K$ dose matrix. When designing a drug-combination trial, one must decide whether to look for a single MTD or multiple MTDs. In some settings, it can be advantageous to find multiple MTDs so that we can further study which one yields the highest synergistic treatment effect. We begin this chapter with designs that look for a single MTD, and we finish with designs that target multiple MTDs.

Before describing the designs, we establish some notation. Let π_{jk} denote the DLT probability for dose combination A_jB_k, and let ϕ denote the target DLT probability for MTD. We use n_{jk} to denote the number of patients who have been assigned to A_jB_k, and y_{jk} to denote the number of DLTs observed at A_jB_k, $j = 1,\ldots,J$ and $k = 1,\ldots,K$. Therefore, at a particular point during the trial, the observed data are $D = \{D_{jk}, j = 1,\ldots,J, k = 1,\ldots,K\}$, where $D_{jk} = (n_{jk}, y_{jk})$ are the data observed at A_jB_k.

4.2   Model-based designs

Numerous model-based designs have been proposed to find a single MTD for combination trials. Thall et al. (2003) proposed a Bayesian drug-combination dose-finding method based on a six-parameter model, assuming that doses are continuous and can be freely changed during the trial. Yin and Yuan (2009a) and Yin and Yuan (2009c) proposed Bayesian dose-finding designs based on a copula-type model and latent contingency tables, respectively. Braun and Wang (2010) developed a dose-finding method based on a Bayesian hierarchical model. Wages et al. (2011) proposed the partial ordering continuous reassessment method (POCRM) by reducing the two-dimensional dose-searching space to a one-dimensional searching line based on partial ordering of the dose combinations of two drugs. Braun and Jia (2013) generalized CRM to handle drug-combination trials. Riviere et al. (2014) proposed a Bayesian dose-finding design based on the logistic model. Cai et al. (2014) and Riviere et al. (2015b) proposed Bayesian adaptive designs for drug-combination trials involving molecularly targeted agents. Neuenschwander et al. (2015) proposed a drug-combination design based on the Bayesian logistic regression model (BLRM) with an overdose control rule. Albeit different, most of these designs adopt a common dose-finding strategy:

1. Devise a model to describe the dose–toxicity surface.

2. Based on the accumulating data, continuously update the model estimate, and make the decision of dose assignment for the incoming new patient, typically by assigning the new patient to the dose whose estimated DLT probability is closest to the target ϕ.

3. Stop the trial when the maximum sample size is reached, and select the MTD based on the estimates of the DLT probabilities.

In what follows, we use the copula-type design (Yin and Yuan, 2009a) as an example to illustrate the core dose-finding strategy of the model-based designs. Let p_j be the prespecified toxicity probability corresponding to dose

A_j of agent A, $p_1 < \cdots < p_J$, and q_k be that of dose B_k of agent B, $q_1 < \cdots < q_K$. Before two drugs are combined, typically they have been individually studied in phase I trials. Thus, the values of p_j's and q_k's can be elicited from clinicians, based on the historical data from previous single-agent trials. The DLT probability of $A_j B_k$ can be modeled using the copula-type regression model (Yin and Yuan, 2009a; Yin and Lin, 2015) as follows:

$$\pi_{jk} = 1 - \{(1 - p_j^\alpha)^{-\gamma} + (1 - q_k^\beta)^{-\gamma} - 1\}^{-1/\gamma}, \tag{4.1}$$

for $j = 1, \ldots, J$ and $k = 1, \ldots, K$, where $\alpha, \beta, \gamma > 0$ are unknown model parameters. An attractive feature of model (4.1) is that if only one drug is tested, it reduces to the power model used in CRM. For example, if there is only drug A (i.e., $q_k = 0$), the model reduces to

$$\pi_j = p_j^\alpha, \quad j = 1, \ldots, J.$$

Therefore, the copula-type regression model can be viewed as a generalization of the single-agent CRM to drug-combination trials. This unique feature allows us to easily incorporate prior information of single-agent trials into the combination trials by specifying p_j's and q_k's, which are known as "skeletons" in the CRM.

Alternatively, one can also use the logistic regression model for π_{jk} (Riviere et al., 2014),

$$\text{logit}(\pi_{jk}) = \beta_0 + \beta_1 u_j + \beta_2 v_k + \beta_3 u_j v_k, \tag{4.2}$$

where β_0, β_1, β_2, and β_3 are unknown parameters, with $\beta_1 > 0$ and $\beta_2 > 0$ to ensure that the toxicity probability is increasing with the increasing dose level of each agent alone, and for all $k, \beta_1 + \beta_3 v_k > 0$ and for all $j, \beta_2 + \beta_3 u_j > 0$ to ensure that the toxicity probability is increasing with the increasing dose levels of both agents together. The covariates u_j and v_k are the standardized doses of A_j and B_k, respectively, defined as $u_j = \log\{p_j/(1 - p_j)\}$ and $v_k = \log\{q_k/(1 - q_k)\}$, to incorporate the prior information on single-agent toxicity probabilities.

Denote $\boldsymbol{\theta}$ as the set of unknown parameters of the dose–toxicity model. For example, $\boldsymbol{\theta} = (\alpha, \beta, \gamma)^T$ in the copula-type regression model, and $\boldsymbol{\theta} = (\beta_0, \beta_1, \beta_2, \beta_3)^T$ in the logistic regression model. Suppose that at a certain stage of the trial, among n_{jk} patients treated at $A_j B_k$, y_{jk} subjects have experienced DLT. The likelihood given the observed data D is

$$L(D \mid \boldsymbol{\theta}) \propto \prod_{j=1}^{J} \prod_{k=1}^{K} \pi_{jk}^{y_{jk}} (1 - \pi_{jk})^{n_{jk} - y_{jk}}.$$

The posterior distribution of $\boldsymbol{\theta}$ is given by

$$f(\boldsymbol{\theta} \mid D) \propto L(D \mid \boldsymbol{\theta}) f(\boldsymbol{\theta}),$$

where $f(\boldsymbol{\theta})$ is the prior distribution of $\boldsymbol{\theta}$. For example, $f(\boldsymbol{\theta}) = f(\alpha)f(\beta)f(\gamma)$ in the copula-type regression model, where $f(\alpha)$, $f(\beta)$, and $f(\gamma)$ denote independent, vague gamma prior distributions with mean one and large variances for α, β, and γ, respectively.

Based on the posterior distribution of the model parameters, the dose-finding algorithm can be described as follows:

1. The first cohort of patients is treated at the lowest dose combination A_1B_1.

2. Suppose the current dose combination is A_jB_k, to determine the dose for the next cohort, consider the following:

    (i) If $\Pr(\pi_{jk} < \phi|D) > c_e$, where c_e is the fixed probability cut-off for dose escalation, the dose for the next cohort of patients moves to an adjacent dose combination chosen from $\{A_{j+1}B_k, A_{j+1}B_{k-1}, A_{j-1}B_{k+1}, A_jB_{k+1}\}$, which has a DLT probability higher than the current doses and closest to ϕ. If the current dose combination is A_JB_K, the dose stays at the same dose combination.

    (ii) If $\Pr(\pi_{jk} < \phi|D) < c_d$, where c_d is the fixed probability cutoff for dose de-escalation, the dose moves to an adjacent dose combination chosen from $\{A_{j-1}B_k, A_{j-1}B_{k+1}, A_{j+1}B_{k-1}, A_jB_{k-1}\}$, which has a DLT probability lower than the current doses and closest to ϕ. If the current dose combination is A_1B_1, the trial is terminated.

    (iii) Otherwise, the next cohort of patients continues to be treated at the current dose combination.

3. Repeat Step 2 until the maximum sample size is reached, and select MTD as the dose whose estimate of DLT probability is closest to ϕ.

The model-based designs perform reasonably well, but for several reasons they are rarely used in practice. First, these designs are statistically and computationally complicated, leading many practitioners to perceive that decisions of dose allocation arise from a "black box," which limits its application in practice. Secondly, robustness is another potential issue for the model-based drug-combination designs. Since these designs use a strategy akin to CRM, one might expect them to share the similar robustness (e.g., consistent under misspecified models (Shen and O'Quigley, 1996)). Unfortunately, that generally is not the case. The consistency of CRM under misspecified models requires several assumptions (Shen and O'Quigley, 1996). A critical one is monotonicity (i.e., the DLT probability monotonically increases with the dose), which does not hold for drug combinations. Based on our experience, model-based drug-combination trial designs are substantially more delicate, and it is not difficult to find scenarios where such designs do not perform well.

4.3 Model-assisted designs

Model-assisted designs have emerged as an attractive approach for phase I drug-combination trials. These designs yield competitive performance similar to that of model-based designs, but are much simpler to implement and more robust without assuming any parametric model on the dose–toxicity surface. This section introduces two model-assisted designs: the BOIN combination design and the keyboard combination design.

4.3.1 BOIN combination design

The BOIN combination design (Lin and Yin, 2017a) uses the same dose escalation/de-escalation rule as the BOIN single-agent design, described in Section 3.4. Let $\hat{\pi}_{jk} = y_{jk}/n_{jk}$ denote the observed DLT rate at $A_j B_k$, and λ_e and λ_d denote the optimal dose escalation and de-escalation boundaries, respectively, as defined in Section 3.4. The BOIN dose escalation/de-escalation rule is: if $\hat{\pi}_{jk} \leq \lambda_e$, escalate the dose; if $\hat{\pi}_{jk} > \lambda_d$, de-escalate the dose; otherwise, retain the current dose.

For combination trials, the new challenge is that there is more than one option for dose escalation: we can escalate the dose of A or the dose of B. Similarly, when we decide to de-escalate the dose, we can de-escalate the dose of A or the dose of B. The BOIN combination design uses the Bayesian posterior probability of $\pi_{jk} \in (\lambda_e, \lambda_d)$ to make the decision. As the posterior distribution of π_{jk} (i.e., beta distribution) is continuous, $\pi_{jk} \in (\lambda_e, \lambda_d) = \pi_{jk} \in (\lambda_e, \lambda_d]$ and thus we use the former throughout. Specifically, define admissible dose escalation and de-escalation sets as

$$\mathcal{A}_E = \{A_{j+1}B_k, A_j B_{k+1}\} \text{ and } \mathcal{A}_D = \{A_{j-1}B_k, A_j B_{k-1}\}.$$

When the BOIN rule says escalate, we escalate to the dose combination that belongs to $\mathcal{A}_E$ and has the highest value of $\Pr(\pi_{jk} \in (\lambda_e, \lambda_d)|D_{jk})$; and when the BOIN rule says de-escalate, we de-escalate to the dose combination that belongs to $\mathcal{A}_D$ and has the highest value of $\Pr(\pi_{jk} \in (\lambda_e, \lambda_d)|D_{jk})$. That is, we always move toward the dose that is most likely to be in the acceptable (or "stay") interval (λ_e, λ_d). The value of $\Pr(\pi_{jk} \in (\lambda_e, \lambda_d)|D_{jk})$ can be easily evaluated based on the beta-binomial model

$$y_{jk} \mid n_{jk}, \pi_{jk} \sim \text{Binomial}(n_{jk}, \pi_{jk}) \tag{4.3}$$
$$\pi_{jk} \sim \text{Beta}(1, 1) \equiv \text{Unif}(0, 1)$$

with the posterior $\pi_{jk} \mid D_{jk} \sim \text{Beta}(y_{jk} + 1, n_{jk} - y_{jk} + 1)$. The BOIN combination design is summarized in Table 4.1.

Because $\Pr\{\pi_{jk} \in (\lambda_e, \lambda_d)|D_{jk}\}$ can be pre-determined for all possible outcomes $D_{jk} = (n_{jk}, y_{jk})$, the dose escalation and de-escalation rule in Step

TABLE 4.1: BOIN drug-combination design.

(a) Patients in the first cohort are treated at the lowest dose combination A_1B_1 or a prespecified dose combination.

(b) Suppose the current cohort is treated at dose combination A_jB_k; to assign a dose to the next cohort of patients:

- If $\hat{\pi}_{jk} \leq \lambda_e$, we escalate the dose to the combination that belongs to $\mathcal{A}_E$ and has the largest value of $\Pr\{\pi_{j'k'} \in (\lambda_e, \lambda_d)|D_{j'k'}\}$. If the current dose combination is A_JB_K, then we retain this dose for treating the next cohort of patients.

- If $\hat{\pi}_{jk} > \lambda_d$, we de-escalate the dose to the combination that belongs to $\mathcal{A}_D$ and has the largest value of $\Pr\{\pi_{j'k'} \in (\lambda_e, \lambda_d)|D_{j'k'}\}$. If the current dose combination is A_1B_1, then we retain this dose for treating the next cohort of patients.

- Otherwise, if $\lambda_e < \hat{\pi}_{jk} \leq \lambda_d$, then the dose stays at the same combination A_jB_k.

(c) Repeat Step (b) until the maximum sample size N is reached, and select MTD as the dose combination whose isotonic estimate (Bril et al., 1984b) of the DLT probability is closest to ϕ.

(b) can be tabulated before the trial begins, which makes the BOIN combination design easy to implement in practice. One important characteristic of the dose transition rule of the BOIN combination design is that, to make the decision, what really is needed is the ordering of $\Pr\{\pi_{jk} \in (\lambda_e, \lambda_d)|D_{jk}\}$ for doses within $\mathcal{A}_E$ and $\mathcal{A}_D$, not their absolute values. Thus, for the purpose of decision making, we only need to tabulate the rank of each possible D_{jk} according to the value of $\Pr\{\pi_{jk} \in (\lambda_e, \lambda_d)|D_{jk}\}$. This greatly simplifies the decision table. We refer to the rank of a dose with D_{jk} as the desirability score of that dose. Table 4.2 shows the desirability score for n_{jk} up to 12, with the cohort size of 3 and the target DLT probability $\phi = 0.3$. A larger value indicates a more desirable dose with a larger value of $\Pr\{\pi_{jk} \in (\lambda_e, \lambda_d)|D_{jk}\}$. To conduct the trial, there is no need for any model fitting or complicated calculation (as required by model-based designs). Users simply look up the desirability score table to make the dose escalation/de-escalation decision.

To illustrate the use of the decision table, suppose that at the current dose A_1B_1, we observed $\hat{\pi}_{11} = 1/6 < \lambda_e = 0.236$ and thus need to escalate the dose. Assume that at this point, the observed data at A_2B_1 and A_1B_2 are $D_{21} = (0,0)$ and $D_{12} = (3,1)$, respectively. To determine which dose the trial should be escalated to, we simply look up Table 4.2 and identify that the desirability scores of A_2B_1 and A_1B_2 are 25 and 40, respectively. As A_1B_2

TABLE 4.2: Desirability score table for the BOIN combination design with the target $\phi = 0.3$. A larger value indicates a higher desirability.

No. Pts	No. DLTs	Des. Score	No. Pts	No. DLTs	Des. Score	No. Pts	No. DLTs	Des. Score
0	0	25	6	3	34	12	0	11
3	0	28	6	≥ 4	E	12	1	22
3	1	40	9	0	14	12	2	45
3	2	24	9	1	32	12	3	61
3	≥ 3	E	9	2	53	12	4	62
6	0	19	9	3	57	12	5	48
6	1	42	9	4	41	12	6	30
6	2	49	9	≥ 5	E	12	≥ 7	E

Note: "No. Pts" is the total number of patients treated, "No. DLTs" is the number of patients who experienced DLT, "Des. Score" is the desirability score. "E" indicates that the dose level is too toxic and should be eliminated from the trial. If both dose levels should be eliminated, then the trial should stay at the current dose level.

has a higher desirability score, the next cohort of patients will be treated at A_1B_2. In the case that A_2B_1 and A_1B_2 have the same desirability scores (e.g., both doses have not yet been used to treat any patients, or have been used to treat patients and generated same data), we can choose one dose randomly or based on other clinical considerations. The BOIN combination decision table, such as Table 4.2, can be generated using the software described later, and included in the trial protocol for trial conduct.

Like the BOIN single-agent design, during the trial conduct the BOIN combination design imposes the dose elimination/early stopping rule such that: if $\Pr(\pi_{jk} > \phi \mid D_{jk}) > 0.95$ and $n_{jk} \geq 3$, dose combination A_jB_k and the higher combinations (i.e., $\{A_{j'}B_{k'}, j' \geq j, k' \geq k\}$), are eliminated from the trial, and the trial is terminated if the lowest dose combination is eliminated, where $\Pr(\pi_{11} > \phi \mid D_{11})$ is evaluated based on the posterior distribution (3.2). The letter "E" in Table 4.2 reflects this dose elimination rule such that a dose with a desirability score of "E" is excessively toxic and should not be considered in the admissible escalation/de-escalation set.

The BOIN combination design employs the same optimal escalation and de-escalation boundaries as the BOIN single-agent design to guide the dose transition, thus it inherits the latter's desirable statistical and operational properties. It is simple, transparent, easy-to-calibrate, and efficient to identify MTD. In addition, because no parametric assumption is made on the dose–toxicity surface, the BOIN combination design is more robust than model-based designs. Extensive simulation studies have demonstrated that the BOIN combination design yields competitive performance comparable to more complicated model-based designs, see Section 4.4 for more details.

4.3.2 Keyboard combination design

The keyboard combination design (Pan et al., 2020) provides another model-assisted design for finding MTD for drug-combination trials. This design uses the same dose escalation/de-escalation rule as the keyboard single-agent design, described in Section 3.3 (e.g., Figure 3.1). That is, let $\mathcal{I}^* = (\delta_1, \delta_2)$ denote the target key (i.e., the pre-specified proper dosing interval), and $\mathcal{I}_{\max}$ denote the strongest key (i.e., the interval with the highest posterior probability, given D_{jk}). If $\mathcal{I}_{\max}$ is located on the left side of $\mathcal{I}^*$ (denoted as $\mathcal{I}_{\max} \prec \mathcal{I}^*$), escalate the dose; if $\mathcal{I}_{\max}$ is located on the right side of $\mathcal{I}^*$ (denoted as $\mathcal{I}_{\max} \succ \mathcal{I}^*$), de-escalate the dose; if $\mathcal{I}_{\max}$ is $\mathcal{I}^*$ (denoted $\mathcal{I}_{\max} \equiv \mathcal{I}^*$), retain the current dose.

Again, for combination trials, the new challenge is that there are two options for dose escalation and de-escalation: escalate the dose of A or the dose of B when $A_j B_k$ is deemed an underdose, or de-escalate the dose of A or the dose of B when $A_j B_k$ is deemed excessively toxic. The keyboard combination design adopts a similar strategy to the BOIN combination design, using $\Pr\{\pi_{jk} \in \mathcal{I}^* | D_{jk}\}$ to determine which drug's dose should be escalated or de-escalated. That is, when the dose escalation is needed, we escalate to the dose combination that belongs to $\mathcal{A}_E$ and has the highest value of $\Pr(\pi_{jk} \in \mathcal{I}^* | D_{jk})$; and when the dose de-escalation is needed, we escalate to the dose combination that belongs to $\mathcal{A}_D$ and has the highest value of $\Pr(\pi_{jk} \in \mathcal{I}^* | D_{jk})$. The keyboard combination design is summarized in Table 4.3.

4.3.3 Waterfall design

The BOIN and keyboard combination designs described above focus on finding a single MTD. For some combination trials, it is of interest to find multiple MTDs (i.e., the MTD contour) from the dose matrix. These MTDs can be further evaluated in subsequent phases of cohort expansion or phase II trials to identify the optimal combination that generates the highest synergistic treatment effect. This section introduces a model-assisted design, the waterfall design (Zhang and Yuan, 2016), which can be used to find the MTD contour.

Finding the MTD contour is substantially more challenging than finding a single MTD. This is because, in order to find all MTDs in the dose matrix, we must explore the whole dose matrix using the limited sample size that is a key characteristic of phase I trials. If we do not explore the whole dose matrix, we risk missing some MTDs. To address this challenge, the waterfall design employs the divide-and-conquer strategy: it partitions the dose matrix into blocks, and then it applies the BOIN single-agent design to each of the blocks to find all MTDs. Each block is known as a "subtrial," and the doses within a block must be fully ordered in toxicity. In other words, the divide-and-conquer strategy reduces the two-dimensional searching space to several

TABLE 4.3: Keyboard drug-combination design.

(a) Patients in the first cohort are treated at the lowest dose combination $A_1 B_1$ or a prespecified dose combination.

(b) Suppose the current cohort is treated at dose combination $A_j B_k$, given the observed data $D_{jk} = (n_{jk}, y_{jk})$, we identify the strongest key $\mathcal{I}_{\max}$ based on the posterior distribution of π_{jk} and assign a dose to the next cohort of patients as follows:

- If $\mathcal{I}_{\max} \prec \mathcal{I}^*$, we escalate the dose to the combination that belongs to $\mathcal{A}_E$ and has the largest value of $\Pr\{\pi_{j'k'} \in \mathcal{I}^* | D j' k'\}$. If the current dose combination is $A_J B_K$, then we retain this dose for treating the next cohort of patients.

- If $\mathcal{I}_{\max} \succ \mathcal{I}^*$, we de-escalate the dose to the combination that belongs to $\mathcal{A}_D$ and has the largest value of $\Pr\{\pi_{j'k'} \in \mathcal{I}^* | D j' k'\}$. If the current dose combination is $A_1 B_1$, then we retain this dose for treating the next cohort of patients.

- Otherwise, if $\mathcal{I}_{\max} \equiv \mathcal{I}^*$, then the dose stays at the same combination $A_j B_k$.

(c) Repeat Step (b) continued until the maximum sample size N is reached, and select the MTD as the dose combination whose isotonic estimate of the DLT probability is closest to ϕ.

---

one-dimensional searching lines (i.e., subtrials) such that the BOIN single-agent design can be directly applied.

As illustrated in Figure 4.2, the waterfall design partitions the $J \times K$ dose matrix into J subtrials (or blocks), within which the doses are fully ordered. Without loss of generality, we assume that $J \leq K$. These subtrials are conducted sequentially from the top of the matrix to the bottom, which is where the design gets its waterfall name. The goal of the waterfall design is to find the MTD contour, which is equivalent to finding MTD in each row of the dose matrix, if one exists. The waterfall design can be described as follows:

1. Divide the $J \times K$ dose matrix into J subtrials $S_J, \cdots, S_1$, according to the dose level of drug A:

$$S_J = \{A_1 B_1, A_2 B_1, \cdots, A_J B_1, A_J B_2, \cdots, A_J B_K\},$$
$$S_{J-1} = \{A_{J-1} B_2, A_{J-1} B_3, \cdots, A_{J-1} B_K\},$$
$$S_{J-2} = \{A_{J-2} B_2, A_{J-2} B_3, \cdots, A_{J-2} B_K\},$$
$$\cdots$$
$$S_1 = \{A_1 B_2, A_1 B_3, \cdots, A_1 B_K\}.$$

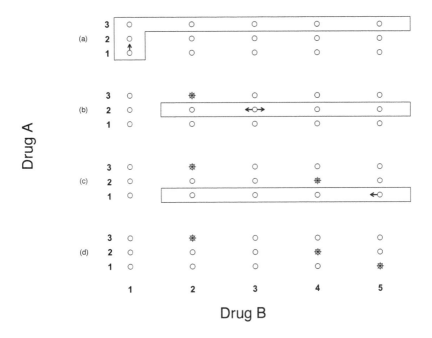

FIGURE 4.2: Illustration of the waterfall design for a 3 × 5 dose matrix. The outlined doses in each panel form a subtrial, and the asterisk denotes the candidate MTD. As shown in panel (a), the first subtrial starts with dose combination $A_1 B_1$. After the first subtrial identifies $A_3 B_2$ as the candidate MTD, the second subtrial starts with dose combination $A_2 B_3$ (see panel (b)). After the second subtrial identifies $A_2 B_4$ as the candidate MTD, the third subtrial starts with dose combination $A_1 B_5$ (see panel (c)). After all subtrials complete, the MTD in each row of the dose matrix is selected based on the data from all subtrials, as shown in panel (d).

> To ensure the trial starts at the lowest dose, subtrial S_J includes lead-in dose combinations $A_1 B_1, A_2 B_1, \cdots, A_J B_1$ (i.e., the first column of the dose matrix, see panel (a) in Figure 4.2). Within each subtrial, the dose combinations are fully ordered with monotonically increasing DLT probabilities.

2. Conduct the subtrials sequentially using the BOIN design (or another single-agent dose-finding method) as follows:

   (i) Conduct subtrial S_J, starting from the lowest dose combination $A_1 B_1$, to find the J-th candidate MTD. We call the dose selected by a particular subtrial the "candidate MTD" to highlight that it may not be selected as an MTD upon completion of the trial, which will be based on the data collected from all the subtrials. The candidate MTD is used to determine which subtrial to conduct next and which combination to assign first.

(ii) If the current subtrial $S_j, j = J, \ldots, 2$ selects $A_{j^*}B_{k^*}$ as the candidate MTD, then conduct subtrial S_{j^*-1} beginning with $A_{j^*-1}B_{k^*+1}$. That is, the next subtrial corresponds to the dose of drug A one level below the candidate MTD from the previous subtrial. Upon completion of subtrial S_{j^*-1}, and selection of another candidate MTD, this rule is applied similarly to determine the next subtrial and its starting dose (e.g., see Figure 4.2).

(iii) Repeat Step (ii) until subtrial S_1 completes.

3. Estimate the DLT probabilities (i.e., π_{jk}), based on the data from all the subtrials using the two-dimensional isotonic regression (Bril et al., 1984b). For each row of the dose matrix, select MTD as the dose combination with an estimated DLT probability closest to the target, unless every dose combination in this row is too toxic.

In Step 2 (ii), the reason subtrial S_{j^*-1} starts with dose $A_{j^*-1}B_{k^*+1}$ rather than the lowest dose in this subtrial (i.e., $A_{j^*-1}B_2$) is that—assuming toxicity monotonically increases in each dose and the candidate MTD in the previous subtrial is an actual MTD—$A_{j^*-1}B_{k^*+1}$ is the lowest dose that could be MTD in subtrial S_{j^*-1}. Therefore, starting from $A_{j^*-1}B_{k^*+1}$ allows the design to quickly find MTD in subtrial S_{j^*-1}. Using Figure 4.2 as an example, dose combination A_3B_2 is the candidate MTD identified in the first subtrial S_3, and thus, the second subtrial S_2 starts with A_2B_3. It is illogical for subtrial S_2 to start with the lowest dose A_2B_2, since the partial ordering implies that A_2B_2 is farther below the MTD contour. Starting S_2 with A_2B_2 in this example would expose excessive numbers of patients to putatively subtherapeutic dose combinations, thereby wasting the already limited patient resources. The waterfall design is efficient because it uses the results from the current subtrial to inform the design of subsequent subtrials.

Because the dose combinations in each subtrial are fully ordered, applying the BOIN single-agent design to them is straightforward. The main challenge is how to determine when the current subtrial should be terminated. One simple approach is to prespecify a maximum sample size for each subtrial. Once the maximum sample size is reached, we stop the subtrial, determine its candidate MTD, and initiate the next subtrial. Although this approach works well for single-agent dose-finding trials, it is inefficient for multiple subtrials. This is because—depending on the number of dose combinations between the starting dose combination and the actual MTD, as well as the shape of the dose–toxicity curve—often each subtrial requires different numbers of patients to identify MTD reliably. Based on this consideration, Zhang and Yuan (2016) proposed and recommended the following "convergence" stopping rule for the subtrials:

If the number of patients treated at the current dose combination reaches or exceeds a prespecified number, say m, then stop the subtrial, select the candidate MTD, and initiate the next subtrial.

The rationale for the stopping rule is that when the patient allocation concentrates at a particular dose combination, this indicates that the dose-finding algorithm likely has converged on the MTD, so we should stop the subtrial and select the current dose combination as the candidate MTD. This stopping rule automatically adjusts the sample size of each subtrial to reflect the difficulty of the dose finding (e.g., the number of dose combinations between the starting dose combination and the actual MTD, and the shape of the dose–toxicity curve). In addition, this stopping rule ensures that a certain number of patients are treated at the candidate MTD, which is achieved in single-agent dose-finding designs using cohort expansion after selecting MTD. Setting $m \geq 9$ ($m = 12$ is preferred) usually ensures reasonable operating characteristics.

Although the above stopping rule provides an automatic, reasonable way to determine the sample size for a particular subtrial, in some cases, it is advantageous to prespecify a maximum sample size for each subtrial as well. This can be done by adding an extra stopping rule:

> If the number of patients treated in subtrial S_j reaches or exceeds n_j^{max}, where n_j^{max} is the prespecified maximum sample size for the subtrial S_j, then stop the subtrial, select the candidate MTD, and initiate the next subtrial.

We recommended setting n_j^{max} between $4\times$(the number of doses in subtrial S_j) and $6\times$(the number of doses in subtrial S_j) for $j = 1, \ldots, J$. For example, a trial with a 3×5 dose matrix, like the trial depicted in Figure 4.2, consists of a first subtrial with seven doses, and second and third subtrials each with four doses. We may set $n_j^{max} = 28$, 16, and 16 for three subtrials, respectively, and thus a maximum of 60 patients for the trial. Although a maximum sample size of 60 patients may seem large, because there are 15 dose combinations, 60 patients actually is not a very large sample size. Consider a single-agent dose-finding trial with 15 doses, the maximum sample size under the 3+3 design is $6 \times 15 = 90$ patients. We recommend using computer simulations to calibrate m and n_j^{max}, thereby ensuring the design has desirable operating characteristics. This simulation-based calibration can be carried out with the software described in the next section.

Lastly, the partition of the dose matrix is not unique. Any partition can be used as long as it is clinically sound and the doses within each block are fully ordered. For example, we may use each row of the dose matrix as a subtrial. In addition to the advantage of being simple and transparent, as shown in the next section, the waterfall design also has competitive and often better performance than the more complicated model-based designs, such as the design based on the product of independent beta probabilities (PIPE) (Mander and Sweeting, 2015).

TABLE 4.4: Average computation time (minutes) needed to simulate 2000 trials for each design to study drug combinations.

Combinations	BLRM	POCRM	Copula	BOIN	PIPE	Waterfall
2×4	192.7	12.2	140.0	0.06	2.4	0.9
3×5	456.7	16.2	302.0	0.07	9.5	0.9
4×4	498.6	16.9	352.1	0.08	11.8	1.0

4.4   Operating characteristics

Liu et al. (2022) performed a comprehensive simulation study to compare the operating characteristics of model-based designs (including the copula design, BLRM, and POCRM) with the model-assisted BOIN combination design. The target DLT probability is $\phi = 25\%$. To avoid cherry-picking scenarios, this study randomly generated 3000 scenarios for each of the 2×4, 3×5, and 4×4 dose combination matrices, and simulated 500 trials under each scenario for each design. The sample size $N = 27$ for 2×4 dose combinations, and $N = 48$ for 3×5, and 4×4 dose combinations, with the cohort size of three. Detailed simulation configurations can be found in Liu et al. (2022).

Figure 4.3 shows the percentage of correct selection of MTD (i.e., a metric for accuracy) and the percentage of patients treated at the toxic doses with true DLT probabilities $\geq 33\%$ (i.e., a metric for safety). Compared to the model-based designs, the BOIN combination design yields the highest percentage of correct selection of MTD, and comparable or better safety. In addition, because of the simplicity of the BOIN combination design, its computational time is just a fraction of that of the model-based designs (see Table 4.4). More importantly, it is significantly simpler to implement the BOIN combination design in practice by using its decision table as described in Section 4.3.1.

Liu et al. (2022) also compared the waterfall design with the model-based PIPE design for finding the MTD contour based on 3000 random scenarios of the 2×4 combination trial with 36 patients, a 3×5 combination trial with 48 patients, and a 4×4 combination trial with 48 patients. The results (see Figure 4.4) support the superior performance of the model-assisted design.

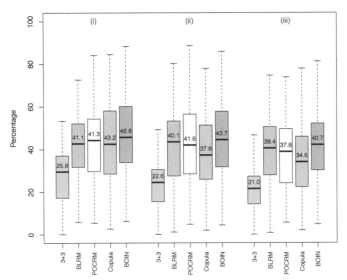

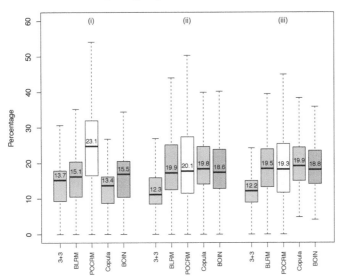

FIGURE 4.3: Simulation results of designs for finding one MTD based on 3000 random scenarios of (i) the 2×4 combination trial with 27 patients, (ii) the 3×5 combination trial with 48 patients, and (iii) the 4×4 combination trial with 48 patients.

Percentage of selecting at least one MTD

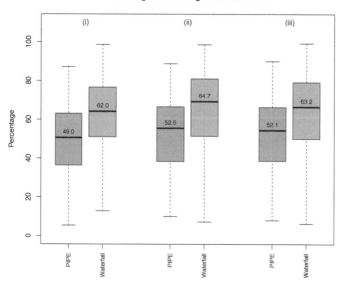

Percentage of patients treated at overdoses

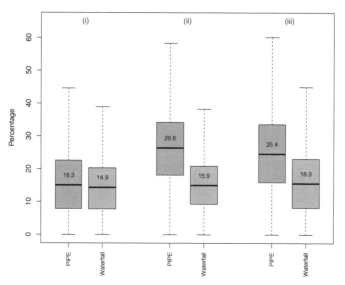

FIGURE 4.4: Simulation results of the PIPE and waterfall designs for finding the MTD contour based on 3000 random scenarios of (i) the 2×4 combination trial with 27 patients, (ii) the 3×5 combination trial with 48 patients, and (iii) the 4×4 combination trial with 48 patients.

4.5 Software and case study

Software

The keyboard combination design can be implemented using the web application "Keyboard Suite," freely available at `http://www.trialdesign.org`. Software for the BOIN combination design is available in several different forms: (1) the web application "BOIN Suite" with intuitive graphical user interface, freely available at `http://www.trialdesign.org` (Figure 4.5 shows the launchpad of BOIN Suite); (2) the R package `BOIN` freely available at CRAN; and (3) the Windows desktop program "BOIN Suite," freely available at the MD Anderson Cancer Center Software Download Website `https://biostatistics.mdanderson.org/softwaredownload/`. We recommend the web application as it is most frequently updated.

Case study

Acute Myeloid Leukemia (AML) Trial   The objective of this phase I trial (ClinicalTrials.gov Identifier: NCT03600155) was to determine MTD for nivolumab and ipilimumab alone and in combination in patients with high risk or refractory/relapsed AML and myelodysplastic syndrome (MDS) following allogeneic stem cell transplantation. The trial consisted of three parallel arms: nivolumab alone, ipilimumab alone, and the combination of nivolumab and ipilimumab. We here focus on the combination arm, which employed the BOIN combination design to find MTD. Three doses of ipilimumab and two doses of nivolumab were investigated (i.e., six combinations) with the target DLT probability $\phi = 30\%$.

We use this trial as an example to illustrate the use of the BOIN web application to design drug-combination trials. After selecting and launching the "BOIN Comb" module from the BOIN Suite launchpad (see Figure 4.5), the trial can be designed using the following three steps:

Step 1: Enter trial parameters

Doses and Target   As shown in Figure 4.6, drug A (nivolumab) has two doses and drug B (ipilimumab) has three doses. The starting dose is A_1B_1. For trials where the lowest dose is believed to be safe, starting from a slightly higher dose level (e.g., A_1B_2) may save the sample size as it allows the design to reach the MTD sooner.

The target DLT probability is $\phi = 0.3$, which should be elicited from clinicians. The target ϕ can be adjusted if a specific safety requirement is desirable, see Section 3.6 for more guidance on the specification of ϕ_1 and ϕ_2. Depending on the trial objective, users can select to find a single MTD (using the BOIN combination design) or the MTD contour (using the waterfall design). In general, finding the MTD contour requires a larger sample size than find a single MTD. This is because finding the MTD contour requires a more thorough

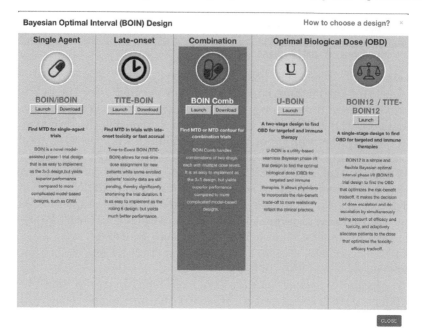

FIGURE 4.5: Launchpad of BOIN Suite.

exploration of the dose matrix. When target $\phi = 0.25$, the software provides an option called "Apply the 3+3 design run-in" to embed the 3+3 design rule into the BOIN decision table, see Section 3.6 for details.

Sample Size and Cohort Size   The cohort size for the trial is three and the number of cohorts is 10, with a total sample size of 30 (Figure 4.7). As a rule of thumb, for a $J \times K$ dose matrix, we recommend the maximum sample size $N \in [4 \times J \times K, 6 \times J \times K]$, where $6 \times J \times K$ corresponds to the maximum sample size of the 3+3 design for $J \times K$ doses. This sample size generally yields reasonable operating characteristics. In the AML trial, $J = 2$ and $K = 3$, thus, the recommended sample size is 24 to 36. To reduce the actual sample size, it is often useful to use the "convergence" stopping rule: stop the trial early when m patients have been assigned to a dose and the decision is to stay at that dose. This stopping criterion suggests that the dose finding approximately converges to the MTD, thus the trial can be stopped. We recommend $m = 9$ or larger. In this trial, $m = 12$ is used. Because of the early stopping rule, the actual sample size used in the trial is often smaller than N. The saving depends on the true dose–toxicity relationship and can be evaluated using simulation.

Another useful approach to reducing the sample size is to conduct the accelerated titration before treating patients in cohorts of three (i.e., checking

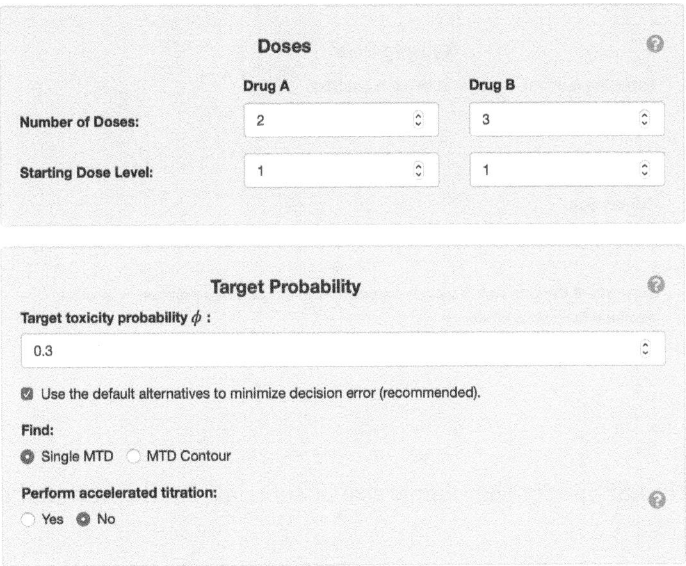

FIGURE 4.6: Specify doses, target, and accelerated titration.

"Yes" for "Perform accelerated titration" in Figure 4.6). During the accelerated titration, we treat patients in cohorts of one, and we continue escalating the dose in the one-patient-per-dose-level fashion until any of the following events is observed: (i) the first instance of DLT, (ii) the second instance of

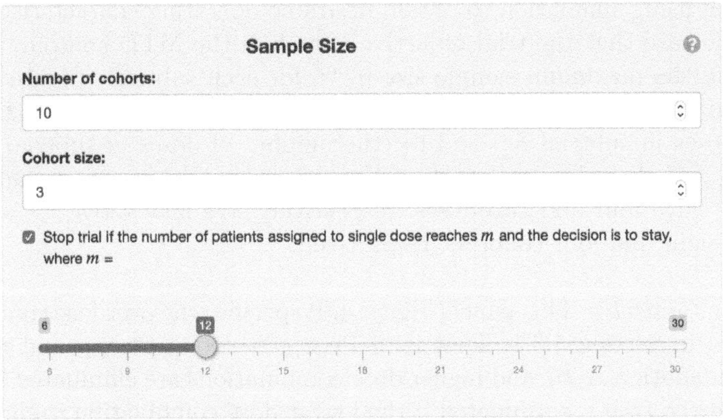

FIGURE 4.7: Specify the sample size, cohort size, and convergence stopping rule.

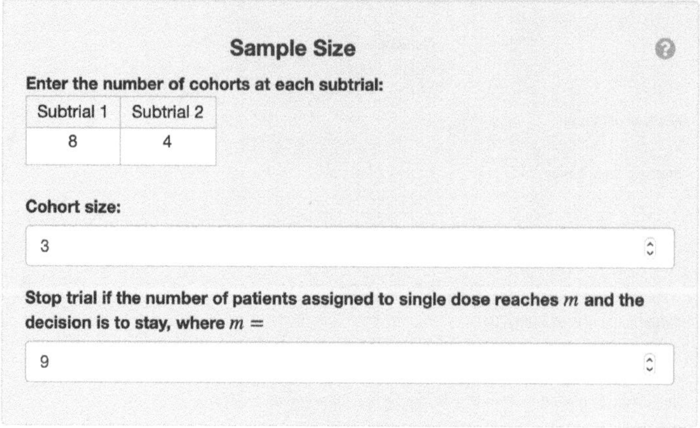

Sample Size   ❔

Enter the number of cohorts at each subtrial:

Subtrial 1	Subtrial 2
8	4

Cohort size:

3	⭥

Stop trial if the number of patients assigned to single dose reaches m and the decision is to stay, where $m =$

9	⭥

FIGURE 4.8: Specify the sample size of subtrials for the waterfall design.

moderate (grade 2) toxicity, or (iii) the highest dose level is reached. At that point, the titration ends. We add two more patients to the current dose level, and hereafter switch to the cohort size of three. During the titration, the drug to be escalated (A or B) can be randomly selected, which is adopted by the software for simulation, or chosen based on clinical consideration. In addition, as described in Section 3.6, in the simulation, the software considers only (i) and (iii), ignoring (ii), due to practical considerations described previously.

As the accelerated titration leads to more aggressive dose escalation, it should be used only when there is sufficient evidence that low doses are most likely underdosing. When the accelerated titration is used, the sample size can be chosen using simulation to obtain desirable operating characteristics.

In the case that the trial objective is to find the MTD contour, we need to specify the maximum sample size n_j^{max} for each subtrial (Figure 4.8). As discussed in Section 4.3.3, we recommend setting n_j^{max} between 4×(the number of doses in subtrial S_j) and 6×(the number of doses in subtrial S_j), for $j = 1, \ldots, J$. For this trial with a 2 × 3 dose matrix, the first and second subtrials have four and two doses, respectively. We may set $n_j^{max} = 24$ and 12 (i.e., eight and four cohorts), respectively.

Overdose Control   This panel (Figure 4.9) specifies the overdose control rule, described in Section 4.3.1. That is, if $\Pr(\pi_{jk} > \phi \,|\, D_{jk}) > p_E$ and $n_{jk} \geq 3$, dose combination $A_j B_k$ and higher dose combinations are eliminated from the trial, and the trial is terminated if the lowest dose combination is eliminated. When the trial is terminated due to toxicity, no dose combination should be selected as MTD. In general, we recommend to use the default probability cutoff $p_E = 0.95$. A smaller value (e.g., 0.9), results in stronger overdose control, but at the cost of reducing the probability of correctly identifying

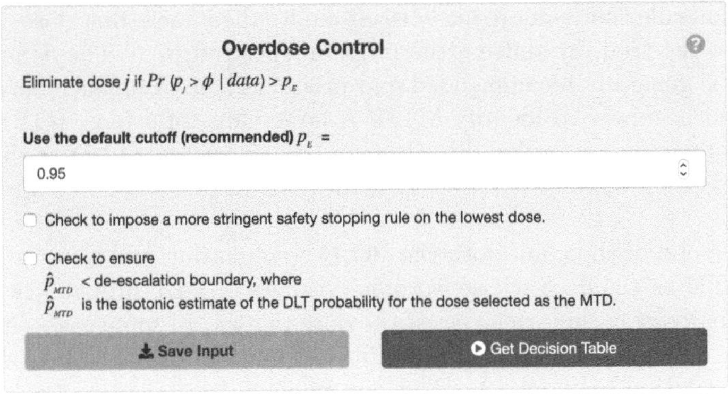

FIGURE 4.9: Specify the overdose control rules.

MTD. This is because, in order to correctly identify MTD, it is imperative to explore the doses sufficiently to learn their toxicity profile.

In some trial settings, under the null case that all doses are overly toxic (i.e., the lowest/first dose combination is above MTD), the probability of early trial termination may not be as high as we desire (e.g., $> 70\%$). That is simply because the small sample size cannot provide enough power to distinguish whether a dose is overly toxic or not (e.g., 40% vs. 30%). To achieve a high early termination probability (when the first dose is overly toxic), we can activate the first option "□ Check to impose a more stringent safety stopping rule" (see Figure 4.10). This option allows the user to impose a more strict dose

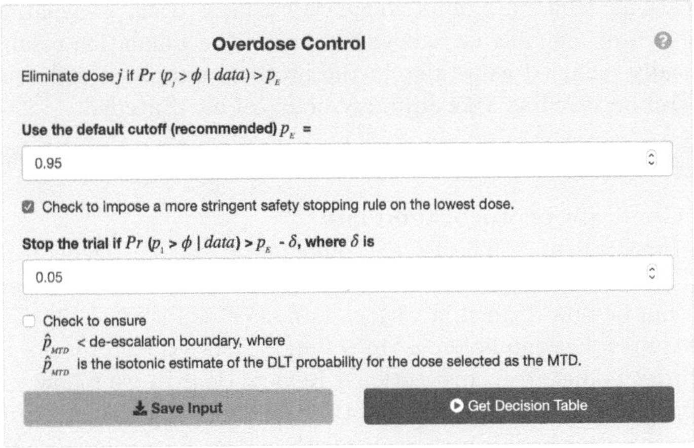

FIGURE 4.10: Adjust the overdose control cutoff.

elimination rule that is more likely to eliminate the lowest/first dose, and thus terminate the trial, by shifting the probability cutoff by δ. The default value $\delta = 0.05$ is generally recommended and produces a good balance between the safety and accuracy to identify MTD. A large value of δ (e.g., 0.1) increases the early termination probability (when all doses are toxic), but at the cost of reducing the probability of correctly identifying MTD (when one of the doses is MTD).

At the end of the trial, both the BOIN combination and waterfall designs select MTD as the dose whose isotonic estimate of DLT probability is closet to ϕ, globally or within each row of the dose matrix. In some cases, it may be desirable to require the DLT probability estimate of MTD to be lower than the de-escalation boundary λ_d. This can be done by activating the option at the bottom of the "Overdose Control" panel (Figure 4.9).

After completing the specification of trial parameters, the decision table will be generated by clicking the "Get Decision Table" button. The decision table and the desirability score table will be automatically included in the protocol template in Step 3, but they also can be saved as a separate csv, Excel, or pdf file in this step if needed.

Step 2: Run simulation

Operating Characteristics   This step generates the operating characteristics of the BOIN combination design or the waterfall design through simulation, see Figure 4.11. The simulation scenarios should cover various possible clinical scenarios (e.g., MTDs are located at different dose combinations). To facilitate the generation of the operating characteristics of the design, the software automatically provides a set of randomly generated scenarios with various MTD locations, which are often adequate for most trials. Depending on the application, users can add or remove scenarios. The simulation results will be automatically included as a table in the protocol template in the next step, but can also be saved as a separate csv or Excel file if needed.

Step 3: Generate protocol template

Protocol Preparation   The BOIN combination software generates sample texts and a protocol template to facilitate the protocol write-up. The protocol template can be downloaded in various formats (see Figure 4.12). Use of this module requires the completion of Steps 1 and 2. Once the protocol is approved by regulatory bodies (e.g., Institutional Review Board), we follow the design decision table as well as the desirability score table to conduct the trial and make adaptive decisions (e.g., dose escalation/stay/de-escalation). When the trial completes, use the "Select MTD" module to select MTD.

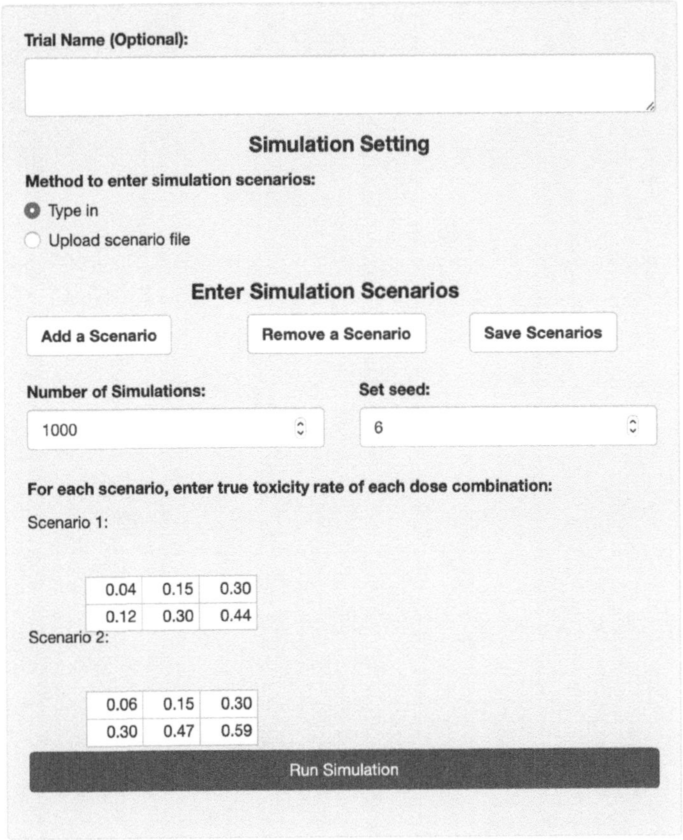

Trial Name (Optional):

Simulation Setting

Method to enter simulation scenarios:

○ Type in

○ Upload scenario file

Enter Simulation Scenarios

| Add a Scenario | Remove a Scenario | Save Scenarios |

Number of Simulations: **Set seed:**

| 1000 | ⌃⌄ | | 6 | ⌃⌄ |

For each scenario, enter true toxicity rate of each dose combination:

Scenario 1:

| 0.04 | 0.15 | 0.30 |
| 0.12 | 0.30 | 0.44 |

Scenario 2:

| 0.06 | 0.15 | 0.30 |
| 0.30 | 0.47 | 0.59 |

Run Simulation

FIGURE 4.11: Simulate operating characteristics of the design.

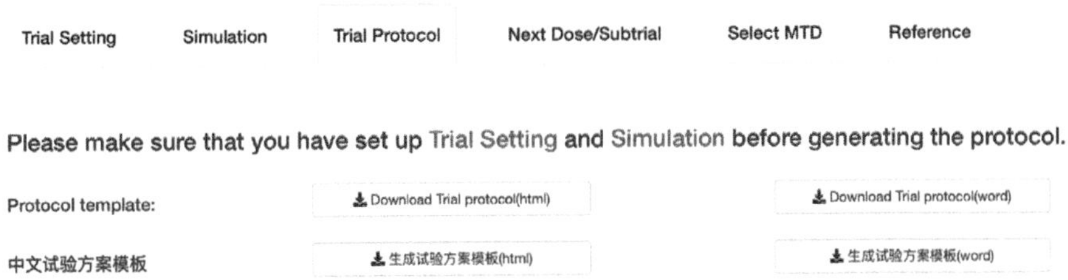

Trial Setting   Simulation   Trial Protocol   Next Dose/Subtrial   Select MTD   Reference

Please make sure that you have set up Trial Setting and Simulation **before generating the protocol.**

Protocol template: ⬇ Download Trial protocol(html)   ⬇ Download Trial protocol(word)

中文试验方案模板   ⬇ 生成试验方案模板(html)   ⬇ 生成试验方案模板(word)

FIGURE 4.12: Download protocol templates.

5

Late-Onset Toxicity

5.1 A common logistical problem

The paradigm for phase I clinical trial design was initially established in the era of cytotoxic chemotherapies, for which toxicities were often acute and ascertainable in the first cycle of therapy. Over the past decade, non-cytotoxic therapies such as molecularly targeted therapies and immunotherapies have entered the clinic. Toxicity associated with these agents is often late onset (Postel-Vinay et al., 2011; Weber et al., 2015), as is that associated with conventional radiotherapy, which may occur several months post-treatment. To account for late-onset toxicity, it is imperative to use a relatively long toxicity assessment window (e.g., over multiple treatment cycles) to define the dose-limiting toxicity (DLT) such that all DLTs relevant to the dose escalation/de-escalation and determination of the maximum tolerated dose (MTD) are captured. This, however, causes a common logistical problem with sequentially outcome-adaptive dose-finding designs, which require the outcomes of previously treated patients to be observed soon enough to apply design's decision rules to choose treatments for new patients. For example, if the DLT takes up to eight weeks to evaluate and the accrual rate is one patient/week, then on average, seven new patients will be accrued while waiting to evaluate the outcome of the first patient. The question is: how can new patients receive timely treatment when the previous patients' outcomes are pending?

The same logistical difficulty arises with rapid accrual. Suppose that DLT of a new agent can be assessed in the first 28-day cycle; if the accrual rate is eight patients/28 days, then on average seven new patients will be accured while waiting to evaluate the first patient's outcome, and the clinicians must determine how to provide them with timely treatment.

This problem, often known as the late-onset (or delayed) outcome problem, persists throughout the trial. Several empirical approaches have been used in practice to alleviate this issue, but they suffer from various flaws. One approach is to turn away new patients and treat them off protocol. This may be less desirable than giving patients the experimental regimen, or impossible if no alternative treatment exists.

Another approach is to give all new patients the dose or treatment that is optimal based on the most recent data. This can have disastrous consequences if the most recently chosen dose later turns out to be overly toxic. For example, suppose that the current dose level j has a total of three patients, but all of them are waiting for toxicity evaluation. Motivated by the desire to complete the trial quickly and based on the most recent data, the dose-assignment decision for a new cohort of three patients is to escalate the dose to $j + 1$, as none of the three patients have the DLT so far. If it turns out that all three patients at dose level j experience severe late-onset toxicities, then the three new patients already have been treated at an excessively toxic dose.

A third approach is to suspend accrual after each cohort and wait until the DLT data for the already accrued patients have cleared before enrolling the next new cohort. This approach of repeatedly interrupting accrual, however, is highly undesirable and often infeasible in practice. It delays treatment for new patients and slows down the trial. Additionally, this approach consequently increases the drug development duration, especially when the accrual is hard as is the case in rare diseases.

In this chapter, we describe several novel phase I designs to handle late-onset toxicities. Compared to the above empirical approaches, these designs are statistically and scientifically rigorous and yield substantially better operating characteristics. In what follows, we first characterize the implications of late-onset toxicities, and then introduce and compare model-based designs and model-assisted designs.

5.2   Late-onset toxicities

In a phase I clinical trial, patients are sequentially treated, and each patient is followed for a fixed period of time (i.e., DLT assessment window), say $(0, \tau)$, to assess the toxicity. For the ith patient, let x_i denote the binary DLT outcome,

$$x_i = \begin{cases} 1, & \text{if a DLT is observed within } (0, \tau), \\ 0, & \text{if a DLT is not observed within } (0, \tau), \end{cases}$$

and let t_i denote the time to DLT. For subjects who have experienced DLT, we have $0 \leq t_i \leq \tau$; for those who do not experience DLT during the DLT assessment window, we set $t_i = \infty$. τ should be elicited from clinicians and large enough to capture all DLTs relevant to MTD determination. For many chemotherapies, τ is often taken as the first cycle of the therapy (e.g., 21 or 28 days), whereas for agents expected to induce late-onset toxicity (e.g., some targeted or immunotherapy agents), τ can be several months or longer after treatment.

At an interim decision time, let u_i $(0 \leq u_i \leq \tau)$ denote the actual follow-up time for patient i. If the patient has finished the DLT assessment, then

$u_i = \tau$. Let δ_i indicate whether the toxicity outcome x_i has been ascertained (i.e., $\delta_i = 1$) or is still pending (i.e., $\delta_i = 0$) by the interim decision time. Then it follows that

$$\delta_i = \begin{cases} 1, & \text{if } t_i \leq u_i \text{ or } u_i = \tau < t_i, \\ 0, & \text{if } t_i > u_i \text{ and } u_i < \tau. \end{cases} \tag{5.1}$$

That is, x_i is unobserved for patients who have not yet experienced toxicity $(t_i > u_i)$ and have not been fully followed up $(u_i < \tau)$; and x_i is observed when patients either have experienced DLT $(t_i \leq u_i)$ or have completed the entire assessment $(u_i = \tau)$ without experiencing DLT.

In general, missing data are more likely to occur for patients who would not experience DLT in $(0, \tau)$. This phenomenon is illustrated in Figure 5.1. The following theorem holds:

Theorem 5.1 *The missing data induced by late-onset toxicity or fast accrual are nonignorable with* $\Pr(\delta_i = 0 | x_i = 0) > \Pr(\delta_i = 0 | x_i = 1)$.

Proof: Considering that each patient is fully followed up to an assessment period τ, if $t_i > \tau$, then $x_i = 0$ and if $t_i \leq \tau$, then $x_i = 1$. For a patient who will not experience DLT within the DLT assessment window, the probability

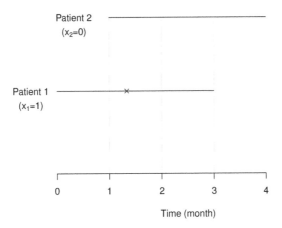

FIGURE 5.1: Illustration of missing outcomes when the toxicity assessment window is three months and accrual rate is one patient per month. For each patient, the horizontal line segment represents the three-month assessment period, during which occurrence of DLT is indicated by a cross. The outcome of patient 1 is missing at month 1, but observed between months 1 and 2. The outcome of patient 2 is missing between months 1 and 4.

that his/her toxicity outcome will be missing is given by

$$
\begin{aligned}
\Pr(\delta_i = 0 \mid x_i = 0) &= \Pr(t_i > u_i, u_i < \tau \mid x_i = 0) \\
&= \Pr(u_i < \tau \mid x_i = 0)\Pr(t_i > u_i \mid u_i < \tau, x_i = 0) \\
&= \Pr(u_i < \tau \mid x_i = 0)\Pr(t_i > u_i \mid u_i < \tau, t_i > \tau) \\
&= \Pr(u_i < \tau),
\end{aligned}
$$

where the last equality follows because t_i and u_i are independent, and $\Pr(t_i > u_i \mid u_i < \tau, t_i > \tau) = 1$. Similarly, for a patient who will experience DLT, the probability that his/her toxicity outcome will be missing is given by

$$
\begin{aligned}
\Pr(\delta_i = 0 \mid x_i = 1) &= \Pr(t_i > u_i, u_i < \tau \mid x_i = 1) \\
&= \Pr(u_i < \tau \mid x_i = 1)\Pr(t_i > u_i \mid u_i < \tau, x_i = 1) \\
&= \Pr(u_i < \tau)\Pr(t_i > u_i \mid u_i < \tau, t_i \le \tau).
\end{aligned}
$$

Because of $\Pr(t_i > u_i \mid u_i < \tau, t_i \le \tau) < 1$, it follows that

$$
\Pr(\delta_i = 0 \mid x_i = 0) > \Pr(\delta_i = 0 \mid x_i = 1).
$$

Therefore, the missing data are more likely to occur for those patients who would not experience DLT in the follow-up period. Because the missing data are nonignorable, the empirical approach that discards the missing data and makes dose-assignment decisions solely based on the observed toxicity data leads to biased inference and poor operating characteristics (Little and Rubin, 2014).

5.3   TITE-CRM

A number of model-based designs have been proposed to address late-onset toxicities. Based on the framework of the continual reassessment method (CRM), Cheung and Chappell (2000) proposed the time-to-event CRM (TITE-CRM) to incorporate pending patients' follow-up times into toxicity evaluation via a weighting scheme. Bekele et al. (2007) proposed to monitor late-onset toxicities using predicted risks. Regarding the late-onset toxicity issue as a missing data problem, Yuan and Yin (2011b) utilized the expectation–maximization algorithm to estimate the dose toxicity probabilities based on the incomplete data to direct dose assignment, and Liu et al. (2013) proposed the Bayesian data-augmentation CRM (DA-CRM) to handle pending toxicity data. Yin et al. (2013) imputed the missing toxicity data based on the Kaplan–Meier estimate. In what follows, we use TITE-CRM as an example to illustrate the model-based approach.

The idea of TITE-CRM is to weigh pending patients by their follow-up proportions when evaluating the likelihood of the CRM model. Recall that CRM assumes a parametric dose–toxicity model

$$\pi_j = q_j^{\exp\{\alpha\}}, \text{ for } j = 1, \ldots, J,$$

where α is the unknown parameter, and $q_1 < \cdots < q_J$ are prior estimates of the DLT probabilities, called the skeleton. Let $\tilde{x}_i$ be the actual observed data at the interim decision-making time, indicating whether patient i has experienced DLT ($\tilde{x}_i = 1$) or not yet ($\tilde{x}_i = 0$) at the moment of the interim decision. It is important to distinguish $\tilde{x}_i$ from x_i (i.e., the final data when patients complete their DLT assessment). Clearly, $\tilde{x}_i = 1$ implies $x_i = 1$, but when $\tilde{x}_i = 0$, x_i can be either 0 or 1 due to late-onset toxicities. Let $d_{[i]} = 1, \ldots, J$ denote the dose level that patient i received. TITE-CRM uses the following weighted likelihood to estimate the toxicity probabilities of the doses:

$$L(D \mid \alpha) = \prod_{i=1}^{n} \left[\omega_i q_{d_{[i]}}^{\exp\{\alpha\}} \right]^{I(\tilde{x}_i=1)} \left[1 - \omega_i q_{d_{[i]}}^{\exp\{\alpha\}} \right]^{I(\tilde{x}_i=0)}, \tag{5.2}$$

where $\omega_i = u_i/\tau$ is the weight and $I(\cdot)$ is an indicator function. Under this weighting scheme, patients who completed the DLT assessment receive a full credit of $\omega_i = 1$, whereas pending patients receive a fractional credit that is proportional to their follow-up time u_i. Combining this weighted likelihood with a prior of α, we obtain the posterior estimate $\hat{\pi}_j$ and the dose escalation and de-escalation decision follows the rule described in Section 2.5.

The specification of $\omega_i = u_i/\tau$ is based on the assumption that the time-to-DLT is uniformly distributed over $(0, \tau)$. In equation 5.2, the individual likelihood for a patient who has experienced DLT before u_i is

$$\begin{aligned}
\Pr(t_i \le u_i) &= \Pr(t_i \le u_i \mid t_i \le \tau) \Pr(t_i \le \tau) \\
&= \omega_i \Pr(x_i = 1) = \omega_i q_{d_{[i]}}^{\exp\{\alpha\}},
\end{aligned}$$

and that for a pending patient with $t_i > u_i$ is

$$\Pr(t_i > u_i) = 1 - \Pr(t_i \le u_i) = 1 - \omega_i q_{d_{[i]}}^{\exp\{\alpha\}}.$$

Cheung and Chappell (2000) considered other more complicated choices of ω_i, and found that the simple uniform weighting scheme $\omega_i = u_i/\tau$ is remarkably robust and works well.

The weighted likelihood used by TITE-CRM is a pseudo-likelihood in the sense that it equals the exact likelihood only when the time to toxicity t_i is uniformly distributed over $(0, \tau)$. A more efficient design is DA-CRM (Liu et al., 2013), which uses the true data likelihood for estimation and decision making. DA-CRM models the time-to-DLT t_i using a flexible piecewise exponential model and uses it to impute the missing x_i based on Bayesian data augmentation. It has been demonstrated that the DA-CRM can treat patients safely and also select MTD with a high probability.

5.4   TITE-BOIN

The time-to-event Bayesian optimal interval (TITE-BOIN) design is an extension of the BOIN design (Section 3.4) to address the issue of late-onset toxicities (Yuan et al., 2018). As a model-assisted design, TITE-BOIN is simple to implement and its decision rule can be pre-tabulated and included in the trial protocol. In addition, it yields outstanding performance comparable to the more complicated model-based TITE-CRM.

5.4.1   Trial design

Suppose that at an interim decision time, a total of n_j patients have been enrolled at the current dose, among which r_j patients have completed the DLT assessment and their DLT data x_i are known, but $c_j = n_j - r_j$ patients have not completed the DLT assessment and their DLT data x_i are pending. Let O_j denote the set of patients whose DLT data are known (i.e., observed with $\delta_i = 1$) and M_j denote the set of patients whose DLT data are pending (i.e., missing with $\delta_i = 0$) at the current dose level j. The estimate of the toxicity rate π_j is given by

$$\hat{\pi}_j = \frac{\sum_{i \in O_j} x_i + \sum_{i \in M_j} x_i}{n_j}.$$

Since the DLT data of patients from the pending set M_j are unknown, we cannot obtain the value of $\hat{\pi}_j$. Without $\hat{\pi}_j$, the dose-assignment rules of BOIN cannot be applied.

TITE-BOIN addresses this issue by imputing the unobserved x_i. Assuming that the time to DLT follows a uniform distribution over $(0, \tau)$, the expected value of x_i, $i \in M_j$, for a pending patient treated at the current dose j with follow-up time u_j is given by

$$
\begin{aligned}
\hat{x}_i &= E(x_i \mid t_i > u_i) \\
&= \Pr(x_i = 1 \mid t_i > u_i) \\
&= \frac{\Pr(x_i = 1) \Pr(t_i > u_i \mid x_i = 1)}{\Pr(x_i = 1) \Pr(t_i > u_i \mid x_i = 1) + \Pr(x_i = 0) \Pr(t_i > u_i \mid x_i = 0)} \\
&= \frac{\pi_j(1 - u_i/\tau)}{\pi_j(1 - u_i/\tau) + (1 - \pi_j)} \\
&\approx \frac{\pi_j(1 - u_i/\tau)}{1 - \pi_j}.
\end{aligned}
$$

Here, the approximation in the last line is appropriate as π_j and thus $\pi_i(1 - u_i/\tau)$ is typically smaller than $1 - \pi_j$. In addition, this approximation slightly inflates the expected value of x_i, which reduces the chance of aggressive dose escalation potentially caused by late-onset toxicity.

By imputing the unobserved x_i, $i \in M_j$, with its expected value $\hat{x}_i$, the toxicity rate estimate $\hat{\pi}_j$ is given by

$$
\begin{aligned}
\hat{\pi}_j &= \frac{\sum_{i \in O_j} x_i + \sum_{i \in M_j} \hat{x}_i}{n_j} \\
&= \frac{\tilde{y}_j + \frac{\pi_j}{1 - \pi_j}(c_j - \text{STFT}_j)}{n_j},
\end{aligned}
\tag{5.3}
$$

where $\tilde{y}_j = \sum_{i \in O_j} x_i$ is the number of DLTs observed so far, $\text{STFT}_j = \sum_{i \in M_j} u_i/\tau$ is the standardized total follow-up time (STFT) for the pending patients at the dose level j, $j = 1, \ldots, J$. For example, given that the DLT assessment window $\tau = 3$ months and, at the current dose, three pending patients have been followed 1, 1.6, and 2.5 months, respectively, the $\text{STFT}_j = (1 + 1.6 + 2.5)/3 = 1.7$. STFT plays a key role in TITE-BOIN for making dose-assignment decisions.

In the statistical literature, the above approach is known as single mean imputation (Little and Rubin, 2014). One drawback of single mean imputation is that, although it provides an unbiased and consistent point estimate, the resulting variance estimate is biased because of ignoring the imputation uncertainty. In our case, this is not a concern as the decision rules of BOIN only rely on the point estimate of π_j. After the single mean imputation, $\hat{\pi}_j$ is a valid point estimate of π_j.

Equation (5.3) involves an unknown value π_j. Yuan et al. (2018) replaced it with its Bayesian posterior mean estimate $\tilde{\pi}_j$ based on the observed data. Assuming the beta-binomial model with the prior $\pi_j \sim \text{Beta}(\alpha, \beta)$, the Bayesian posterior mean estimate is given by $\tilde{\pi}_j = (\tilde{y}_j + \alpha)/(r_j + \alpha + \beta)$. A vague beta prior with $\alpha = \phi/2$ and $\beta = 1 - \alpha$ is recommended such that the prior corresponds to an effective prior sample size of 1 with the prior mean $\phi/2$. As described in Section 5.2, $\tilde{\pi}_j$ is based on the observed data, and thus overestimates π_j. This is actually beneficial because it slightly inflates $\hat{\pi}_j$, reducing the chance of aggressive dose escalation potentially caused by late-onset toxicity.

One important property of equation (5.3) is that $\hat{\pi}_j$ is a monotonically decreasing function of STFT. Therefore, determining whether $\hat{\pi}_j$ crosses the BOIN escalation and de-escalation boundaries λ_e and λ_d (i.e., $\hat{\pi}_j \leq \lambda_e$ or $\hat{\pi}_j > \lambda_d$) is equivalent to determining whether $\text{STFT}_j \geq \Delta_e$ or $\text{STFT}_j < \Delta_d$, with

$$
\begin{aligned}
\Delta_e &= \left\{ c_j - \frac{1 - \tilde{\pi}_j}{\tilde{\pi}_j}(n_j \lambda_e - \tilde{y}_j) \right\} I\left(\frac{\tilde{y}_j}{n_j} < \phi \right) + \infty I\left(\frac{\tilde{y}_j}{n_j} \geq \phi \right) \\
\Delta_d &= \left\{ c_j - \frac{1 - \tilde{\pi}_j}{\tilde{\pi}_j}(n_j \lambda_d - \tilde{y}_j) \right\} I\left(\frac{\tilde{y}_j}{n_j} > \phi \right) - \infty I\left(\frac{\tilde{y}_j}{n_j} \leq \phi \right),
\end{aligned}
$$

for $j = 1, \ldots, J$. Here, the indicator functions $I\left(\frac{\tilde{y}_j}{n_j} < \phi \right)$ and $I\left(\frac{\tilde{y}_j}{n_j} > \phi \right)$ are imposed to ensure that TITE-BOIN has the similar long-memory coherence property as BOIN (i.e., the dose is never escalated/de-escalated if the observed DLT rate $\tilde{y}_j/n_j$ is greater/smaller than the target DLT rate ϕ). As a result, the TITE-BOIN dose escalation/de-escalation rule becomes the following:

- If $\mathrm{STFT}_j \geq \Delta_e$, escalate the dose to the next higher level.

- If $\mathrm{STFT}_j < \Delta_d$, de-escalate the dose to the next lower level.

- Otherwise, stay at the current dose level.

Because the formulas for Δ_e and Δ_d are functions of summary statistics such as n_j, $\tilde{y}_j$, c_j, and STFT_j, this dose escalation/de-escalation rule can be tabulated prior to trial conduct. As a result, TITE-BOIN inherits the transparency and simplicity of the standard BOIN design and does not require repeated, complicated model fitting after treating each cohort/patient.

Table 5.1 shows the TITE-BOIN decision rule with a cohort size of three and a target DLT rate of 0.2. To conduct the trial, we only need to count the number of patients at the current dose, the number of patients who experienced DLT, the number of pending patients, and STFT, and then look up the table to determine the dose escalation/de-escalation. Suppose that three patients have been treated at the current dose, one of them had DLT. We de-escalate the dose regardless of STFT. Consider another case where nine patients have been cumulatively treated at the current dose, one patient had DLT, and four patients have DLT data pending. To treat the next cohort of patients, if STFT of the four pending patients is greater than 2.15, we escalate the dose; otherwise, we retain the current dose. Table 5.1 assumes a cohort size of three, but TITE-BOIN allows any prespecified cohort size, and the corresponding decision table (i.e., similar to Table 5.1 but with more rows) can be easily generated using the software described later.

In principle, TITE-BOIN supports continuous accrual and allows for real-time dose assignment whenever a new patient arrives. To avoid risky decisions caused by sparse data, an accrual suspension rule is imposed:

> If at the current dose, more than 50% of the patients' DLT outcomes are pending, suspend the accrual to wait for more data to become available.

This rule corresponds to "Suspend accrual" in Table 5.1. In addition, the same overdose control/safety stopping rule as BOIN is also employed:

> If $\Pr(\pi_j > \phi \,|\, D_j) > 0.95$ and $n_j \geq 3$, dose level j and higher are eliminated from the trial, and the trial is terminated if the lowest dose level is eliminated.

This overdose control rule corresponds to the decision "Y&Elim," representing "Yes & Eliminate," under the column entitled "De-escalate" in Table 5.1.

TABLE 5.1: Dose escalation and de-escalation rule for TITE-BOIN with a target DLT rate of 0.2 and a cohort size of three, up to 12 patients.

No. treated	No. DLTs	No. data pending	STFT Escalate	STFT Stay	STFT De-escalate
3	0	≤ 1	Y		
3	0	≥ 2		Suspend accrual	
3	1	≤ 2			Y
3	≥ 2	≤ 1			Y&Elim
6	0	≤ 3	Y		
6	0	≥ 4		Suspend accrual	
6	1	≤ 3		Y	
6	1	≥ 4		Suspend accrual	
6	2	≤ 4			Y
6	≥ 3	≤ 3			Y&Elim
9	0	≤ 4	Y		
9	0	≥ 5		Suspend accrual	
9	1	≤ 2	Y		
9	1	3	≥ 0.77	< 0.77	
9	1	4	≥ 2.15	< 2.15	
9	1	≥ 5		Suspend accrual	
9	2	0		Y	
9	2	1		> 0.52	≤ 0.52
9	2	2		> 1.59	≤ 1.59
9	2	3		> 2.66	≤ 2.66
9	2	4		> 3.73	≤ 3.73
9	2	≥ 5		Suspend accrual	
9	3	≤ 6			Y
9	≥ 4	≤ 5			Y&Elim
12	0	≤ 6	Y		
12	0	≥ 7		Suspend accrual	
12	1	≤ 5	Y		
12	1	6	≥ 1.24	< 1.24	
12	1	≥ 7		Suspend accrual	
12	2	≤ 6		Y	
12	2	≥ 7		Suspend accrual	
12	3, 4	≤ 9			Y
12	≥ 5	≤ 7			Y&Elim

Note: "No. treated" is the total number of patients treated at the current dose level, "No. DLTs" is the number of patients who experienced DLT at the current dose level, "No. data pending" denotes that number of patients whose DLT data are pending at the current dose level, "STFT" is the standardized total follow-up time for the patients with data pending, defined as the total follow-up time for the patients with data pending divided by the length of the DLT assessment window. "Y" represents "Yes," and "Y&Elim" represents "Yes & Eliminate." When a dose is eliminated, all higher doses should also be eliminated.

5.4.2   Incorporating prior information

In some trials, prior information is available on the distribution of t_i (i.e., time to DLT). For example, for a certain drug, we may know *a priori* that the DLT is more likely to occur in the later part of the DLT assessment window $(0, \tau)$. Such prior information can be easily incorporated into TITE-BOIN by assuming t_i follows a piecewise uniform prior distribution, which partitions $(0, \tau)$ into several intervals and specifies a uniform distribution within each interval. By increasing the number of partitions, the piecewise uniform distribution can approximate any shape of the time-to-toxicity distribution. For many applications, three partitions (i.e., $(0, \tau/3)$, $(\tau/3, 2\tau/3)$, and $(2\tau/3, 1)$), are often adequate for practical use. Let ν_1, ν_2, and ν_3 ($\nu_1 + \nu_2 + \nu_3 = 1$) be the prior probabilities that the DLT would occur at the three intervals, respectively. Let $h_0 = 0$, $h_1 = \tau/3$, $h_2 = 2\tau/3$, and $h_3 = \tau$, and define

$$\tilde{u}_{ik} = \begin{cases} \tau, & u_i > h_k \\ u_i - h_{k-1}, & u_i \in (h_{k-1}, h_k], \quad k = 1, 2, 3. \\ 0, & \text{otherwise} \end{cases}$$

Then, the conditional probability $\Pr(t_i > u_i \mid x_i = 1)$ is given by

$$\Pr(t_i > u_i \mid x_i = 1) = 1 - \sum_{k=1}^{3} 3\nu_k \tilde{u}_{ik}/\tau.$$

Using similar algebra and approximation, we have

$$\hat{\pi}_j = \frac{\tilde{y}_j + \dfrac{\pi_j}{1 - \pi_j}(c_j - \text{WSTFT}_j)}{n_j}, \tag{5.4}$$

where WSTFT is the weighted STFT, given by

$$\text{WSTFT}_j = \sum_{i \in M_j} \sum_{k=1}^{3} 3\nu_k \tilde{u}_{ik}/\tau.$$

When $\nu_1 = \nu_2 = \nu_3 = 1/3$ (i.e., assigning an equal weight over the DLT assessment window), WSTFT reduces to STFT. In other words, equation (5.3) using the uniform weight is a special case of equation (5.4).

One attractive feature of TITE-BOIN is that using a non-uniform (informative) prior for t_i does not alter its decision table, such as Table 5.1, generated under the non-informative uniform prior. That is, the TITE-BOIN decision table is invariant to the prior distribution of t_i. This is because equation (5.4) has the exact same form as equation (5.3), except that STFT_j is replaced by WSTFT_j. In other words, Table 5.1 (generated by using the non-informative uniform prior) can be used to make the decision of dose escalation/de-escalation, with STFT_j replaced by WSTFT_j, when an informative prior of t_i is used.

5.4.3   Statistical properties

Another desirable feature of TITE-BOIN is that its decision rule is invariant to the length of the assessment window τ, because STFT has been standardized by the latter. This means that given a target DLT rate ϕ, the same decision table can be used to guide dose escalation and de-escalation, regardless of the value of τ. For example, Table 5.1 can be used for any trial with $\phi = 0.2$, regardless of the length of the assessment window. This is practically appealing and greatly simplifies trial protocol preparation. In practice the assessment window varies from one trial to another, depending on the experimental drug and patient population, whereas the target DLT rate is often taken as 0.2, 0.25, or 0.3. Another attractive feature of TITE-BOIN is that when there is no pending DLT data, it seamlessly reduces to the BOIN design.

When dealing with late-onset toxicities, the top concern is patient safety, as the patients with pending outcome data who have not experienced DLT at the interim decision time may yet experience DLT late in the follow-up period. Any reasonable design that handles late-onset toxicity should take that fact into account in its decision making, which can be described by the monotonicity property. Let $D_j = \{(\tilde{x}_i, t_i), i = 1, \ldots, n_j\}$ be the actually observed interim data at dose level j, and let D_j^s denote the "cross-sectional" interim data obtained by setting $x_i = \tilde{x}_i$ (i.e., treating the patients' temporary DLT outcomes at the interim time as their final DLT outcomes at the end of the assessment window). Because of late-onset toxicity, for pending patients, $\tilde{x}_i$ (i.e., the DLT status observed at the interim) is not necessarily equal to x_i (i.e., the DLT status at the end of assessment window). More precisely, $\tilde{x}_i = 1$ implies $x_i = 1$, but when $\tilde{x}_i = 0$, x_i can be either 0 or 1. Let $a(D_j) = -1, 0$ and 1 denote the decisions of dose de-escalation, retaining the current dose, and dose escalation, respectively, based on the data D_j.

DEFINITION (Monotonicity) A dose-finding design is monotonic if $a(D_j) \leq a(D_j^s)$ for $j = 1, \ldots, J$.

Monotonicity indicates that the decision of dose transition based on the observed data D_j should be less aggressive than that based on D_j^s. This is a property that any reasonable design should obey to reflect that patients who have not experienced DLT by the interim decision time may yet experience DLT late in the follow-up period. The TITE-BOIN design has this property because $\hat{x}_i \geq \tilde{x}_i$, thus leading to $\hat{\pi}_j > \tilde{y}_j / n_j$, where $\tilde{y}_j / n_j$ is the observed toxicity rate based on the "cross-sectional" data D_j^s.

Theorem 5.2 *The TITE-BOIN design is monotonic.*

5.4.4   Operating characteristics

Yuan et al. (2018) compared the operating characteristics of the TITE-BOIN design, the 3+3 design, the rolling six design (Skolnik et al., 2008), and the

TABLE 5.2: Dose–toxicity scenarios used in the simulation study. The target dose (i.e., MTD) is bolded.

Scenario	Dose Level						
	1	2	3	4	5	6	7
1	**0.3**	0.4	0.5	0.6	0.7	0.8	0.9
2	0.14	**0.3**	0.39	0.48	0.56	0.64	0.7
3	0.07	**0.23**	0.41	0.49	0.62	0.68	0.73
4	0.05	0.15	**0.3**	0.4	0.5	0.6	0.7
5	0.05	0.12	0.2	**0.3**	0.38	0.49	0.56
6	0.01	0.04	0.08	0.15	**0.3**	0.36	0.43
7	0.02	0.04	0.08	0.1	0.2	**0.3**	0.4
8	0.01	0.03	0.05	0.07	0.09	**0.3**	0.5

TITE-CRM design using simulation studies. Table 5.2 shows eight scenarios used to simulate trial data. The target DLT probability is 30%. The DLT assessment window is three months, the accrual rate is two patients/month. The time to DLT is sampled from a Weibull distribution, with 50% of DLTs occurring in the second half of the assessment window. The maximum sample size is 36 patients, and patients are treated in cohorts of three. Because the 3+3 and rolling six designs often stopped the trial early (e.g., when two of three patients experienced DLT) before reaching 36 patients, in these cases the remaining patients are treated at the selected "MTD" as the cohort expansion, such that the four designs have comparable sample sizes. For the 3+3 design, a new cohort is enrolled only when the previous cohort's DLT data are cleared. More details about the simulation study can be found in Yuan et al. (2018).

Figure 5.2 shows the relative performance of the rolling six, TITE-BOIN and TITE-CRM designs against the performance of the 3+3 design, including the differences in (a) the percentage of correct selection of MTD, (b) the percentage of patients overdosed (i.e., treated at doses above MTD), (c) the percentage of patients underdosed (i.e., treated at doses below the MTD), and (d) the average trial duration. TITE-BOIN is more efficient and has higher percentages of correct selection of MTD than the algorithm-based design (i.e., the rolling six design) because of using the follow-up times of pending patients to determine dose escalation and de-escalation. Compared to the model-based TITE-CRM, TITE-BOIN has comparable accuracy to identify MTD, but it is safer and much simpler. Compared to the 3+3 design, TITE-BOIN dramatically shortens the trial duration owing to its ability to make real-time decisions in the presence of pending patients.

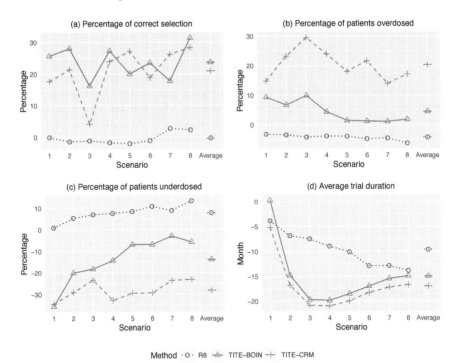

FIGURE 5.2: The relative performance of the rolling six (R6), TITE-BOIN, and TITE-CRM designs against the performance of the 3+3 design, including the differences in (a) the percentage of correct selection of the MTD, (b) the percentage of patients overdosed (i.e., treated at doses above the MTD), (c) the percentage of patients underdosed (i.e., treated at doses below the MTD), and (d) the average trial duration.

5.5 A unified approach using "effective" data

We have discussed the use of mean imputation to address the late-onset toxicity for the BOIN design. That approach, however, cannot be applied to other model-assisted designs such as the mTPI and keyboard designs because these designs make decisions based on the posterior distribution of π_j. In this section, we introduce a more general approach that is applicable to all model-assisted designs (Lin and Yuan, 2020).

In the presence of late-onset toxicity, the interim patients can be divided into two subgroups: the patients with ascertained toxicity outcomes (i.e., $\delta_i = 1$), and the patients with pending outcomes (i.e., $\delta_i = 0$). For a patient with $\delta_i = 1$, we have $\tilde{x}_i = x_i$. Assuming that patient i received dose level j, the likelihood is given by

$$\Pr(\tilde{x}_i = x_i, \delta_i = 1) = \pi_j^{x_i}(1 - \pi_j)^{1-x_i}. \tag{5.5}$$

For a patient with $\delta_i = 0$, x_i has not been ascertained yet with DLT outcome pending, and his/her actual observed outcome is $\tilde{x}_i = 0$. These pending patients are a mixture of two subgroups: patients who will not experience DLT (i.e., $x_i = 0$), and patients who will experience DLT (i.e., $x_i = 1$) but have not experienced it yet by the interim decision time (i.e., $u_i < t_i$). Therefore, the likelihood for a pending patient is given by

$$\begin{aligned}
&\Pr(\tilde{x}_i = 0, \delta_i = 0) \\
=\ & \Pr(x_i = 0)\Pr(\tilde{x}_i = 0, \delta_i = 0 | x_i = 0) + \Pr(x_i = 1)\Pr(\tilde{x}_i = 0, \delta_i = 0 \mid x_i = 1) \\
=\ & \Pr(x_i = 0) + \Pr(x_i = 1)\Pr(t_i > u_i \mid x_i = 1) \\
=\ & 1 - \pi_j + \pi_j\{1 - \Pr(t_i \le u_i \mid x_i = 1)\} \\
=\ & 1 - \pi_j\omega_i,
\end{aligned}$$

where $\omega_i = \Pr(t_i \le u_i \mid x_i = 1)$ can be treated as a weight function, which will be discussed later. Of note, for pending patients, $\tilde{x}_i$ only takes a value of 0 because once $\tilde{x}_i = 1$ (i.e., the patient experiences the DLT), x_i is observed.

Therefore, given the interim data D_j observed at dose level j, the joint likelihood function is given by

$$\begin{aligned}
L(D_j \mid \pi_j) &\propto \prod_{i=1}^{n_j} \pi_j^{\delta_i x_i}(1 - \pi_j)^{\delta_i(1-x_i)}(1 - \omega_i\pi_j)^{1-\delta_i} \\
&= \pi_j^{\tilde{y}_j}(1 - \pi_j)^{m_j} \prod_{i=1}^{n_j}(1 - \omega_i\pi_j)^{1-\delta_i}, \tag{5.6}
\end{aligned}$$

where $\tilde{y}_j = \sum_{i=1}^{n_j} \delta_i x_i$ is the number of patients who experienced DLT by the interim time, and $m_j = \sum_{i=1}^{n_j} \delta_i(1 - x_i) = r_j - \tilde{y}_j$ is the number of patients who have completed the assessment without experiencing DLT.

Let $f(\pi_j)$ denote the prior distribution for π_j (e.g., $f(\pi_j) = \text{Beta}(1,1)$). The posterior distribution $f(\pi_j \mid D_j)$ is given by

$$f(\pi_j \mid D_j) \propto f(\pi_j)L(D_j \mid \pi_j).$$

Recall that without late-onset toxicity, the posterior $f(\pi_j \mid D_j)$ follows a beta distribution, see equation (3.2). In the presence of pending DLT data, $f(\pi_j \mid D_j)$ however does not have a standard distributional form. Although $f(\pi_j \mid D_j)$ can be sampled using the Markov chain Monte Carlo method, it prohibits the enumeration of the decision rule and thus destroys the simplicity of model-assisted designs.

To circumvent this issue and maintain the simplicity of the model-assisted designs, Lin and Yuan (2020) proposed to approximate the last term in (5.6) as follows,

$$(1 - \omega_i\pi_j)^{1-\delta_i} \approx (1 - \pi_j)^{\omega_i(1-\delta_i)}. \tag{5.7}$$

When $\delta_i = 0$, this is a first-order Taylor expansion approximation, noting that the Taylor expansion of $(1 - \pi_j)^{\omega_i}$ at $\pi_j = 0$ is

$$(1 - \pi_j)^{\omega_i} = 1 - \omega_i \pi_j + \omega_i (\omega_i - 1) \pi_j^2 + \cdots .$$

When $\delta_i = 1$, the approximation actually is exact as $(1 - \omega_i \pi_j)^{1-\delta_i} = (1 - \pi_j)^{1-\delta_i} = 1$. Thus, the likelihood (5.6) is approximated as

$$L(D_j \mid \pi_j) \propto \pi_j^{\tilde{y}_j} (1 - \pi_j)^{\tilde{n}_j - \tilde{y}_j}, \tag{5.8}$$

where

$$\tilde{n}_j = \tilde{y}_j + \tilde{m}_j, \quad \tilde{m}_j = m_j + \sum_{i=1}^{n_j} (1 - \delta_i) \omega_i. \tag{5.9}$$

This approximation is simple but extremely powerful. It converts the non-regular likelihood (5.6) into a standard binomial likelihood arising from "effective" binomial data $\tilde{D}_j = (\tilde{n}_j, \tilde{y}_j)$, where $\tilde{n}_j = \tilde{y}_j + \tilde{m}_j$ is the effective sample size, and $\tilde{m}_j$ is the effective number of patients who have not experienced DLT. Because the model-assisted designs in previous chapters are based on the binomial likelihood, these designs can be seamlessly extended to accommodate pending DLT data using the approximated likelihood (5.8), as demonstrated in the next section. In addition, because the approximated likelihood (5.8) depends on the aggregated value of the ω_i's, rather than the individual value of ω_i, the approximation renders it possible to enumerate the decision rules for the resulting design, maintaining the most important feature of the model-assisted designs.

The following theorem establishes that (5.8) provides an accurate approximation of the exact likelihood (5.6).

Theorem 5.3 *Let* $l(\omega_i, \delta_i, \pi_j) = (1 - \omega_i \pi_j)^{1-\delta_i}$ *and* $\tilde{l}(\omega_i, \delta_i, \pi_j) = (1 - \pi_j)^{\omega_i(1-\delta_i)}$. *The approximation error is bounded by*

$$d(\omega_i, \delta_i, \pi_j) = |l(\omega_i, \delta_i, \pi_j) - \tilde{l}(\omega_i, \delta_i, \pi_j)| \leq (1 - \beta_j \pi_j) - (1 - \pi_j)^{\beta_j},$$

where $\beta_j = \log\{-\pi_j / \log(1 - \pi_j)\} / \log(1 - \pi_j)$. *Specifically, for any* $\pi_j \leq 0.4$, *the approximation error* $d(\omega_i, \delta_i, \pi_j) < 0.0255$.

The proof of Theorem 5.3 is provided in Lin and Yuan (2020). To illustrate the accuracy of the approximation, consider two examples: (a) $n_j = 5$ patients have been treated and only $m_j = 2$ patients have finished the assessment without any DLT, and the weights ω_i for the remaining three patients are 0.3, 0.4, and 0.5, respectively, leading to $\tilde{m}_j = 2 + 0.3 + 0.4 + 0.5 = 3.2$; (b) $n_j = 12$ patients have been treated, $\tilde{y}_j = 1$ DLT has been observed, $m_j = 4$ patients have finished the assessment without any DLT, and the weights ω_i for the remaining seven patients are $0.1, 0.2, \ldots, 0.7$, respectively, leading to $\tilde{m}_j = 4 + 0.1 + \cdots + 0.7 = 6.8$. Given the prior $\pi_j \sim \text{Unif}(0, 1)$, Figure 5.3 shows the near coincidence between the exact and approximated posterior distribution functions, indicating that the approximation is quite accurate.

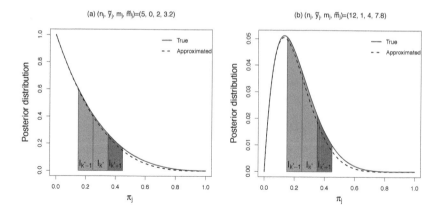

FIGURE 5.3: The exact and approximated posterior functions based on the observed data: (a) $n_j = 5$ patients have been treated and only $m_j = 2$ patients have finished the assessment without any DLT, and the weights for the remaining three patients are 0.3, 0.4, and 0.5; (b) $n_j = 12$ patients have been treated, $\tilde{y}_j = 1$ DLT has been observed, $m_j = 4$ patients have finished the assessment without any DLT, and the weights for the remaining seven patients are $0.1, 0.2, \ldots, 0.7$. The prior distribution of π_j is $\pi_j \sim \text{Unif}(0, 1)$; $\tilde{y}_j$ and $\tilde{m}_j$ represent the "effective" numbers of patients with DLT and patients without DLT, respectively, by the interim time γ; $\mathcal{I}_{k^*}$ represents the target key in the keyboard design.

We now discuss how to specify weight ω_i that appears in the approximated likelihood (5.8). By construction, the weight ω_i adjusts for the fact that the DLT outcome for the pending patient i has not been ascertained yet. Following TITE-CRM and TITE-BOIN, the simplest way is to assume that the time to toxicity t_i is uniformly distributed over the assessment period $(0, \tau)$, leading to

$$\omega_i = \Pr(t_i < u_i \mid x_i = 1) = u_i/\tau. \tag{5.10}$$

As a result, ω_i can be interpreted as the follow-up proportion that patient i has finished. Based on this uniform weight, the effective sample size $\tilde{n}_j$ in the approximated likelihood (5.8) can be easily calculated as

$$
\begin{aligned}
\tilde{n}_j \; &= \; \frac{\text{Total follow-up time for pending patients at dose } j}{\text{Length of assessment window}} \\
&\quad + \text{No. of non-pending patients at dose } j. \\
&= \; \text{STFT}_j + r_j
\end{aligned}
$$

Although the uniform scheme seems very restrictive, it yields remarkably robust performance. This was also observed in TITE-CRM and TITE-BOIN.

As described in section 5.4.2, a more flexible weighting scheme is to assume that t_i follows a piecewise uniform distribution, which partitions $(0, \tau)$ into several intervals and assumes a uniform distribution within each interval. This approach is useful to incorporate prior information on t_i. For ease of exposition, the assessment window $(0, \tau)$ is partitioned into the initial part $(0, \tau/3)$, the middle part $(\tau/3, 2\tau/3)$, and the final part $(2\tau/3, \tau)$, and it assumes that t_i is uniformly distributed in each interval. Let (ν_1, ν_2, ν_3) be the prior probability that the DLT would occur at the three parts of the assessment window, where $\nu_1 + \nu_2 + \nu_3 = 1$. For example, prior data may suggest that the DLT is more likely to occur late in the assessment window, in which case we can choose $\nu_3 > \nu_2 > \nu_1$. It then follows that

$$
\omega_i = \Pr(t_i < u_i \mid x_i = 1) = \begin{cases} 3\nu_1 u_i/\tau, & u_i \in (0, \tau/3), \\ \nu_1 - \nu_2 + 3\nu_2 u_i/\tau, & u_i \in (\tau/3, 2\tau/3), \\ \nu_1 + \nu_2 - 2\nu_3 + 3\nu_3 u_i/\tau, & u_i \in (2\tau/3, \tau), \end{cases}
$$

where ω_i can be interpreted as the weighted follow-up probability that patient i has completed the assessment time. Lin and Yuan (2020) also discussed a more complicated data-driven adaptive weighting scheme, however it only provides negligible improvement over the above two simpler schemes.

5.6 TITE-keyboard and TITE-mTPI designs

Application of the above effective data approach to the keyboard design is straightforward. We refer to the resulting design based on the approximated likelihood (5.8) as the TITE-keyboard design. The decision rule of the TITE-keyboard design is almost the same as that of the keyboard design. The only difference is that to make decisions of dose escalation and de-escalation, the complete data $D_j = (n_j, y_j)$, which are not observable when some DLT data are pending, are replaced by the effective binomial data $\tilde{D}_j = (\tilde{n}_j, \tilde{y}_j)$ for calculating the posterior distribution of π_j and identifying the strongest key. Once the strongest key is identified, the same dose escalation/de-escalation rule is used to guide the dose transition.

Compared to the model-based TITE-CRM, the most appealing feature of TITE-keyboard is that its dose transition rule can be tabulated before the trial begins; see Table 5.3 as the decision table with the target DLT rate $\phi = 0.3$. To conduct the trial, there is no need for real-time model fitting, investigators only need to count the number of patients with DLTs (i.e., $\tilde{y}_j$), the number of patients with data pending (i.e., $c_j = \sum_{i=1}^{n_j}(1-\delta_i)$), the effective number of patients without DLT (i.e., $\tilde{m}_j$), and then use the decision table to determine the dose assignment for the next new cohort. Like the TITE-BOIN design, another feature of the TITE-keyboard design is that its decision

TABLE 5.3: Dose escalation and de-escalation boundaries for TITE-keyboard with a target DLT probability of 0.3 and cohort size of three, up to 12 patients.

n_j	$\tilde{y}_j$	$\tilde{c}_j$	Escalation	Stay	De-escalation
3	0	≤ 1	Y		
3	0	≥ 2		Suspend accrual	
3	1	0		Y	
3	1	$[1,2]$		$\tilde{m}_j > 1.88$	$\tilde{m}_j \leq 1.88$
3	2	≤ 1			Y
3	3	0			Y&Elim
6	0	≤ 6	Y		
6	1	≤ 1	Y		
6	1	$[2,3]$	$\tilde{m}_j \geq 3.07$	$\tilde{m}_j < 3.07$	
6	1	$[4,5]$	$\tilde{m}_j \geq 3.07$	$1.88 < \tilde{m}_j > 3.07$	$\tilde{m}_j \leq 1.88$
6	2	0		Y	
6	2	$[1,4]$		$\tilde{m}_j > 3.75$	$\tilde{m}_j \leq 3.75$
6	3	≤ 3			Y
6	4	≤ 2			Y&Elim
9	0	≤ 9	Y		
9	1	≤ 4	Y		
9	1	$[5,6]$	$\tilde{m}_j \geq 3.07$	$\tilde{m}_j < 3.07$	
9	1	$[7,8]$	$\tilde{m}_j \leq 3.07$	$1.88 < \tilde{m}_j < 3.08$	$\tilde{m}_j \leq 1.88$
9	2	0		Y	
9	2	$[1,3]$	$\tilde{m}_j \geq 6.15$	$\tilde{m}_j < 6.15$	
9	2	$[4,7]$	$\tilde{m}_j \geq 6.15$	$3.75 < \tilde{m}_j < 6.15$	$\tilde{m}_j \leq 3.75$
9	3	0		Y	
9	3	$[1,6]$		$\tilde{m}_j > 5.63$	$\tilde{m}_j \leq 5.63$
9	4	≤ 5			Y
9	5	≤ 4			Y&Elim
12	0	≤ 12	Y		
12	1	≤ 7	Y		
12	1	$[8,9]$	$\tilde{m}_j \geq 3.07$	$\tilde{m}_j < 3.07$	
12	1	$[10,11]$	$\tilde{m}_j \geq 3.07$	$1.88 < \tilde{m}_j < 3.08$	$\tilde{m}_j \leq 1.88$
12	2	≤ 3	Y		
12	2	$[4,6]$	$\tilde{m}_j \geq 6.15$	$\tilde{m}_j < 6.15$	
12	2	$[7,10]$	$\tilde{m}_j \geq 6.15$	$3.75 < \tilde{m}_j < 6.15$	$\tilde{m}_j \leq 3.75$
12	3	≤ 3		Y	
12	3	$[4,9]$		$\tilde{m}_j > 5.63$	$\tilde{m}_j \leq 5.63$
12	4	0		Y	
12	4	$[1,8]$		$\tilde{m}_j > 7.50$	$\tilde{m}_j \leq 7.50$
12	5,6	≤ 7			Y
12	7	≤ 5			Y&Elim

Note: n_j is the number of patients at dose level j, $\tilde{y}_j$ is the number of DLTs observed by the decision time, $\tilde{c}_j = \sum_{i=1}^{n_j}(1 - \delta_i)$ is the number of patients who have data pending, and $\tilde{m}_j$ is the effective number of patients without any DLT. Dose escalation is not allowed if fewer than two patients at dose level j have finished the assessment. "Y" represents "Yes," and "Y&Elim" represents "Yes & Eliminate." When a dose is eliminated, all higher doses should also be eliminated.

table does not depend on the weighting scheme or the length of the DLT assessment window. Table 5.3 applies no matter which of the aforementioned three weighting schemes is used and no matter the length of the assessment window. This is because the likelihood (5.8) depends on $\tilde{m}_j$, which is the effective number of patients without DLT. Moreover, when all the pending DLT data become available, we have $(\tilde{n}_j, \tilde{y}_j) = (n_j, y_j)$. As a result, the TITE-keyboard design reduces to the standard keyboard design in a seamless way.

For patient safety, Lin and Yuan (2020) required that dose escalation is not allowed until at least two patients have completed the DLT assessment at the current dose level. In addition, an overdose control/stopping rule is imposed: at any time during the trial if any dose j satisfies $\Pr(\pi_j > \phi \mid n_j, \tilde{y}_j) > \eta$ and $n_j \geq 3$, then that dose and any higher doses are regarded as overly toxic and should be eliminated from the trial, and the dose is de-escalated to level $j - 1$ for the next cohort of patients, where η is the prespecified elimination cutoff, say $\eta = 0.95$. If the lowest dose level is eliminated, the trial should be terminated early. Table 5.3 also reflects such safety and overdose control rules.

The TITE-keyboard design has some desirable statistical properties. First, similar to the TITE-BOIN design, the TITE-keyboard design also enjoys the monotonicity property.

Theorem 5.4 *The TITE-keyboard design is monotonic.*

Second, the TITE-keyboard design is long-memory coherent. The proofs of Theorems 5.4 and 5.5 are given in Lin and Yuan (2020).

Theorem 5.5 *The TITE-keyboard design is long-memory coherent in the sense that if the empirical toxicity rate $\tilde{\pi}_j = \tilde{y}_j/\tilde{n}_j$ at the current dose is greater (or less) than the target toxicity rate, the design will not escalate (nor de-escalate) the dose, with probability one.*

The effective data approach is general. It is also applicable to BOIN. We compute $\tilde{\pi}_j = \tilde{y}_j/\tilde{n}_j$ using the observed effective data, then the BOIN dose escalation/de-escalation rule can be applied. The effective data approach also can be directly applied to the mTPI design. The only change needed is to replace the complete data (n_j, y_j), potentially unobserved due to late-onset toxicity or fast accrual, with the observed effective data $(\tilde{n}_j, \tilde{y}_j)$ in (3.2) when calculating the unit probability mass (UPM). Once the three UPMs are determined, the decision rules remain the same. We refer to the resulting design as the TITE-mTPI design.

Numerical study shows that TITE-keyboard has excellent performance similar to TITE-BOIN and TITE-CRM. However, TITE-mTPI does not perform as well as the other designs, and it has a high risk of overdosing patients because of the deficiency of mTPI (Section 3.2), see Lin and Yuan (2020) for details of the numerical study.

5.7   Software and case study

Software

The web application for the TITE-BOIN design is available as a module of the BOIN Suite at `http://www.trialdesign.org`. Figure 5.4 shows the launchpad of BOIN Suite. The TITE-keyboard design also can be implemented using the web-based application at `http://www.trialdesign.org`. The two graphical user interface-based software programs allow users to generate the dose-assignment decision table, conduct simulations, obtain the operating characteristics of the design, and generate a trial design template for protocol preparation. The R codes for other time-to-event model-assisted designs are available on Github (`https://github.com/ruitaolin/TITE-MAD`).

Case study

Recurrent or High Grade Gynecologic Cancer Trial   The objective of this phase I trial (ClinicalTrials.gov Identifier: NCT03508570) is to determine

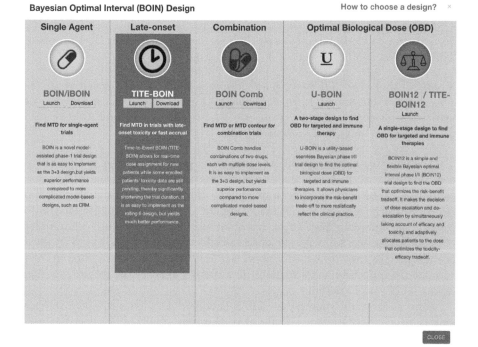

FIGURE 5.4: The launchpad of web application BOIN Suite available at `http://www.trialdesign.org`.

MTD and the recommended phase II dose (RP2D) of intraperitoneal nivolumab in combination with ipilimumab in treating patients with recurrent or high-grade gynecologic cancer. As late-onset toxicities are anticipated for this combination immunotherapy, the DLT assessment window is set as 12 weeks and the TITE-BOIN design is employed to conduct the trial. The maximum sample size is 24 and patients are treated in a cohort size of three. The expected accrual rate is two patients per month. Four doses of ipilimumab will be investigated in combination with a fixed dose of nivolumab, and the target DLT probability is $\phi = 30\%$.

We use this trial as an example to illustrate how to use TITE-BOIN to design phase I trials with late-onset toxicities. After selecting and launching the TITE-BOIN module from the BOIN Suite launchpad (Figure 5.4), the trial is desgined using the following three steps:

Step 1: Enter trial parameters

Doses and Sample Size   As shown in Figure 5.5, the number of doses under investigation is four, starting dose level is one. For trials where the lowest dose is believed to be safe, starting from a slightly higher dose level (e.g., two) may save the sample size as it allows the design to reach MTD sooner. However, due to limited knowledge on the safety of the new drug, in general it is not recommended to start from a high dose level (e.g., four).

The cohort size for the trial is three and the number of cohorts is eight, with a total sample size of 24. When $m = 12$ patients have been assigned to a dose and the decision is to stay at that dose, the trial will be stopped early

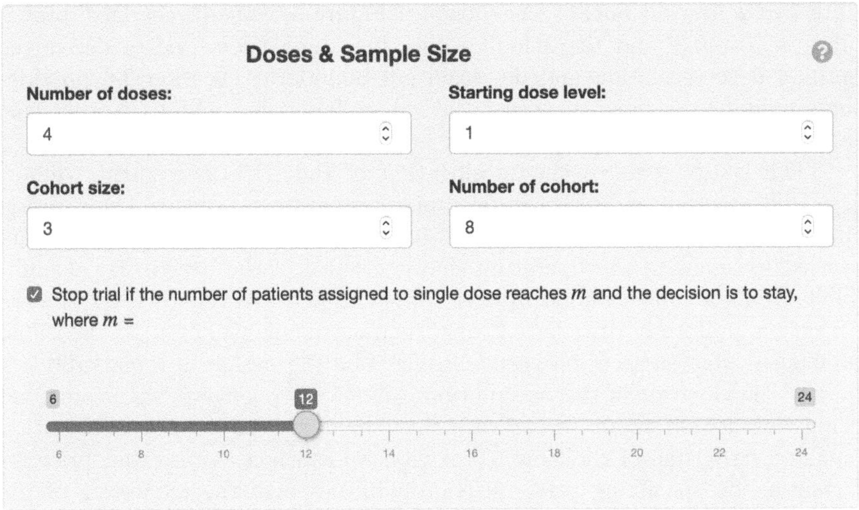

FIGURE 5.5: Specify doses, sample size, and convergence stopping rule.

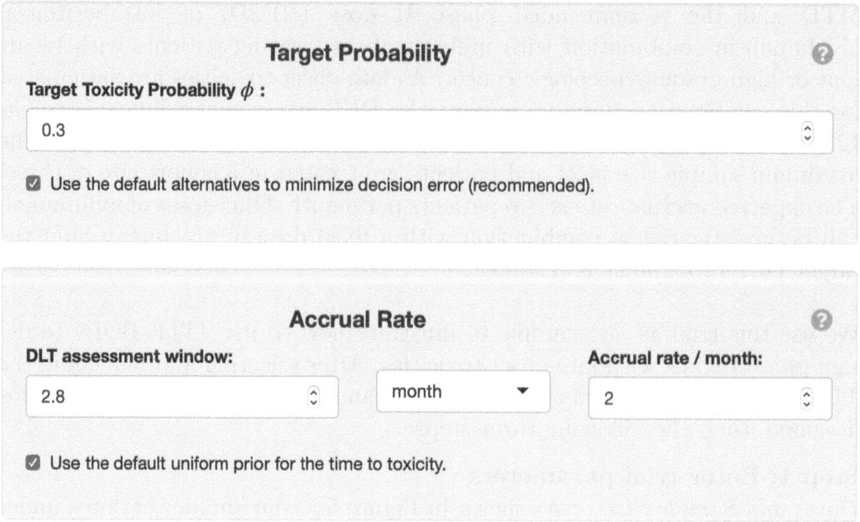

Target Probability   ❓

Target Toxicity Probability ϕ :

| 0.3 | ⌄ |

☑ Use the default alternatives to minimize decision error (recommended).

Accrual Rate   ❓

DLT assessment window: Accrual rate / month:

| 2.8 | ⌄ | month ▾ | 2 | ⌄ |

☑ Use the default uniform prior for the time to toxicity.

FIGURE 5.6: Specify the target, DLT assessment window, and accrual rate.

even if the maximum sample size of 24 is not reached. Because the TITE-BOIN design is a generalization of BOIN, one can follow the rule of BOIN to specify the maximum sample size and the convergence criterion for early stopping, see more details in Section 3.6.

Target and Accrual Rate   As shown in Figure 5.6, the target DLT probability is $\phi = 0.3$, and the default values for (ϕ_1, ϕ_2) are used to derive the optimal dose escalation and de-escalation boundaries. Section 3.6 provides some guidance on how to adjust the values for ϕ, ϕ_1, and ϕ_2 if a specific safety requirement is desirable.

TITE-BOIN requires the specification of the DLT assessment window ($\tau = 2.8$ months) and the accrual rate (two patients/month). The decision rule of TITE-BOIN does not depend on the accrual rate. The accrual rate is used to simulate the operating characteristics of the design. By default, TITE-BOIN assumes a uniform prior for the time to toxicity (i.e., the time to toxicity is uniformly distributed over the assessment window). As described previously, this assumption seems strong, but the design is remarkably robust to the violation of this assumption. Therefore, in general, we recommend using this default prior. In the case that there is reliable prior information on the distribution of the time to toxicity, we can incorporate that prior information by specifying prior DLT probabilities over the trimesters of the assessment window (see Section 5.4.2). For example, Figure 5.7 sets the prior DLT probabilities as 0.3, 0.5, and 0.2 for the trimesters of the assessment window to reflect that DLT is more likely to occur in the middle of the assessment window.

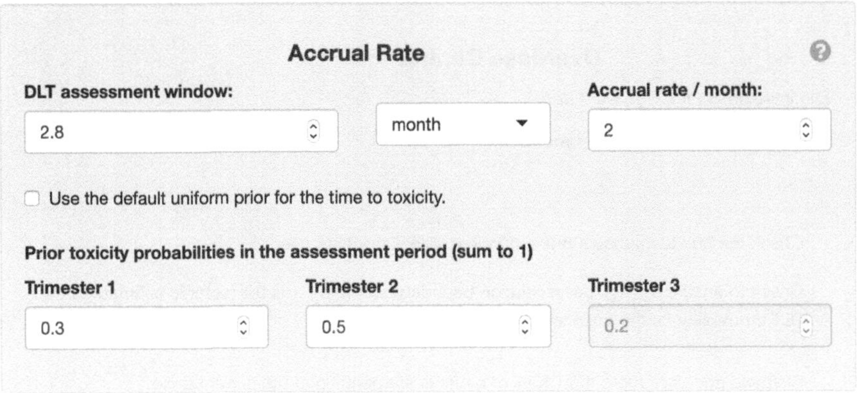

FIGURE 5.7: Specify an informative prior for the time to toxicity.

Overdose Control This panel (see Figure 5.8) specifies the overdose control rule, described in Section 5.4. That is, if $\Pr(\pi_j > \phi \mid D_j) > 0.95$ and $n_j \geq 3$, dose level j and higher are eliminated from the trial, and the trial is terminated if the lowest dose level is eliminated. When the trial is terminated due to toxicity, no dose should be selected as MTD. see Section 3.6 for a detailed discussion on how to specify overdose control through this panel to accommodate various trial objectives.

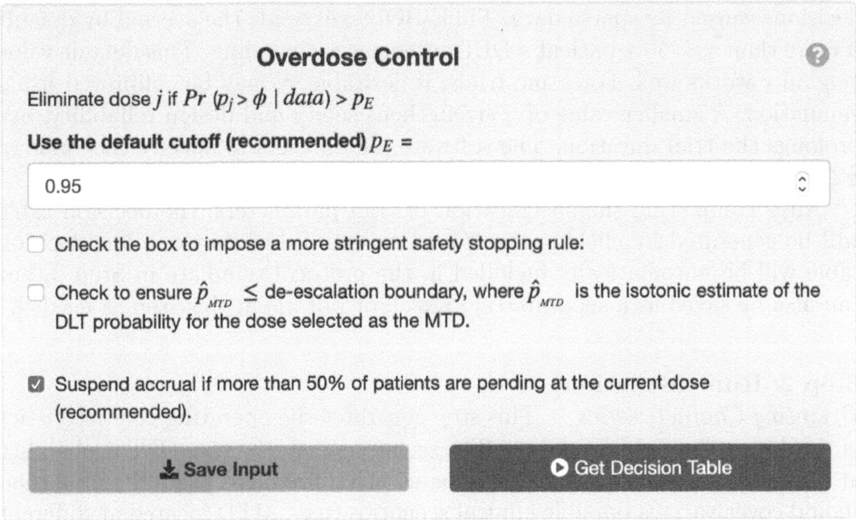

FIGURE 5.8: Specify the overdose control rules.

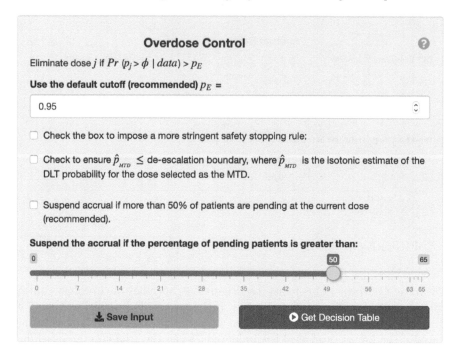

FIGURE 5.9: Adjust the cutoff of the accrual suspension rule.

TITE-BOIN additionally has an accrual suspension rule. To avoid risky decisions caused by sparse data, TITE-BOIN suspends the accrual by default if more than $\gamma = 50\%$ patient's DLT outcomes are pending. This default value generally works well. For some trials, if desirable, γ may be calibrated using simulation. A smaller value of γ strengthens safety and design reliability, but prolongs the trial duration. The software allows users to specify the value of γ (Figure 5.9), but requires $\gamma \leq 65\%$ for safety.

After completing the specification of trial parameters, the decision table will be generated by clicking the "Get Decision Table" button. The decision table will be automatically included in the protocol template in Step 3, but can also be saved as a separate csv, Excel, or pdf file in this step, if needed.

Step 2: Run simulation

Operating Characteristics   This step generates the operating characteristics of the design through simulation. The scenarios (i.e., the true DLT probability at each dose, as shown in the lower panel of Figure 5.10) used for simulation should cover various possible clinical scenarios (e.g., MTD located at different dose levels). To facilitate the generation of the operating characteristics of the design, the software automatically provides a set of randomly generated scenarios with various MTD locations, which are often adequate for most trials. Depending on the application, users can add or remove scenarios. The software provides three distributions to simulate the time to toxicity: the

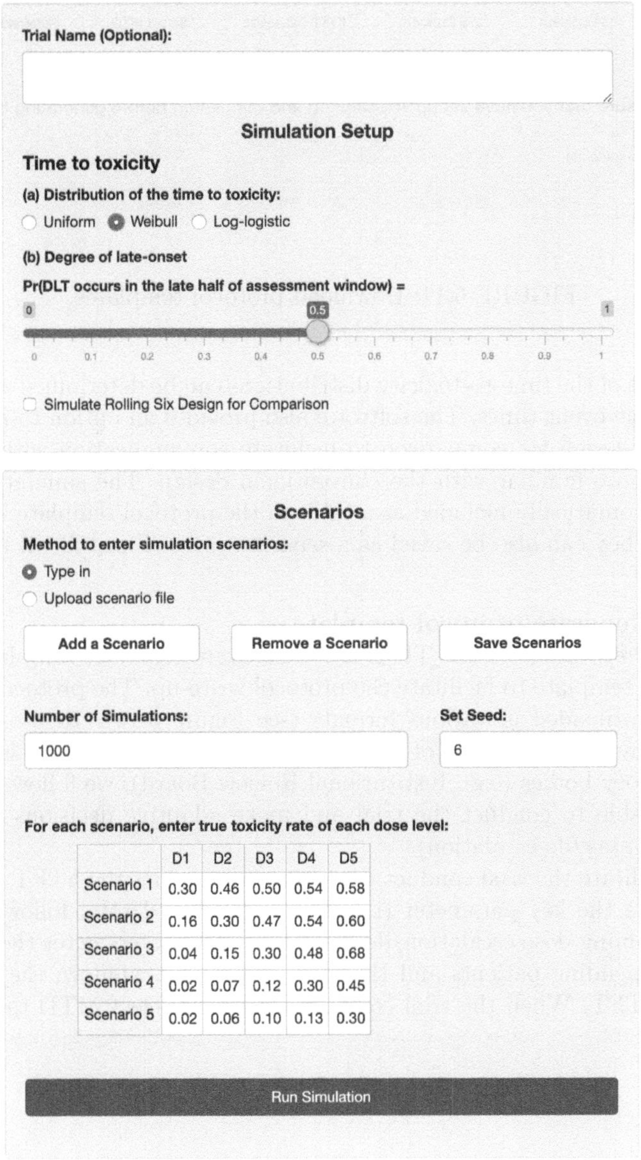

Trial Name (Optional):

Simulation Setup

Time to toxicity

(a) Distribution of the time to toxicity:

○ Uniform   ● Weibull   ○ Log-logistic

(b) Degree of late-onset

Pr(DLT occurs in the late half of assessment window) =

| 0 | 0.5 | 1 |

0   0.1   0.2   0.3   0.4   0.5   0.6   0.7   0.8   0.9   1

☐ Simulate Rolling Six Design for Comparison

Scenarios

Method to enter simulation scenarios:

● Type in
○ Upload scenario file

| Add a Scenario | Remove a Scenario | Save Scenarios |

Number of Simulations: **Set Seed:**

1000   6

For each scenario, enter true toxicity rate of each dose level:

	D1	D2	D3	D4	D5
Scenario 1	0.30	0.46	0.50	0.54	0.58
Scenario 2	0.16	0.30	0.47	0.54	0.60
Scenario 3	0.04	0.15	0.30	0.48	0.68
Scenario 4	0.02	0.07	0.12	0.30	0.45
Scenario 5	0.02	0.06	0.10	0.13	0.30

Run Simulation

FIGURE 5.10: Simulate operating characteristics of the design.

uniform, Weibull, or log-logistic distribution (see Figure 5.10). When the Weibull or log-logistic distribution is selected, users should additionally specify the probability that DLT occurs in the later half of the assessment window, which quantifies the degree of late-onset. With these specifications, the

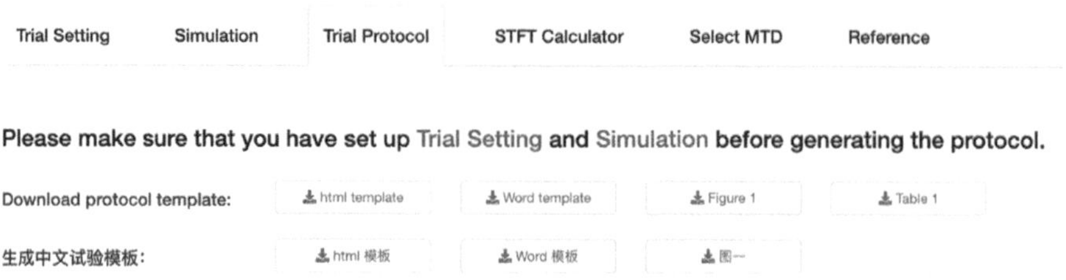

Trial Setting	Simulation	Trial Protocol	STFT Calculator	Select MTD	Reference

Please make sure that you have set up Trial Setting **and** Simulation **before generating the protocol.**

Download protocol template: ⬇ html template   ⬇ Word template   ⬇ Figure 1   ⬇ Table 1

生成中文试验模板：   ⬇ html 模板   ⬇ Word 模板   ⬇ 图一

FIGURE 5.11: Download protocol templates.

parameters of the time-to-toxicity distributions can be determined and used to simulate the event times. The software also provides an option to include the rolling six design for comparison to facilitate communication with clinicians who are more familiar with the conventional design. The simulation results will be automatically included as a table in the protocol template in the next step, but they can also be saved as a separate csv or Excel file, if needed.

Step 3: Generate protocol template

Protocol Preparation   The TITE-BOIN software generates sample texts and a protocol template to facilitate the protocol write-up. The protocol template can be downloaded in various formats (see Figure 5.11). To use this module requires the completion of Steps 1 and 2. Once the protocol is approved by regulatory bodies (e.g., Institutional Review Board), we follow the design decision table to conduct the trial and make adaptive decisions (e.g., dose escalation/stay/de-escalation).

To facilitate the trial conduct, the software provides the STFT Calculator to compute the key parameter (i.e., the standardized total follow-up time), for determining dose escalation/de-escalation. After users enter the follow-up times for pending patients and the DLT assessment window, the calculator outputs STFT. When the trial completes, use the Select MTD tab to select MTD.

6

Incorporating Historical Data

6.1 Historical data and prior information

Incorporating historical data or real-world data (RWD) has great potential to improve the efficiency of phase I clinical trials and to accelerate drug development. The US Food and Drug Administration (FDA) released a draft of guidelines for submitting documents using RWD or evidence to the FDA for drugs and biologics (FDA, 2019).

When designing phase I trials, prior information is often available from previous studies. For example, the drug to be investigated has been studied previously in other indications, or similar drugs belonging to the same class have been studied in earlier phase I trials (Zohar et al., 2011). Phase I bridging trials are another example. These trials extend a drug from one ethnic group (e.g., Caucasian) to another (e.g., Asian) (Liu et al., 2015) or from adult patients to pediatric patients (Petit et al., 2018). In such cases, dose–toxicity data from the original trial in one ethnic group or adult patients can be used to inform the design of the subsequent bridging trials, see for example Liu et al. (2015), Morita (2011), and Li and Yuan (2020).

In this chapter, we first describe how to incorporate prior information in the model-based continual reassessment method (CRM) based on the intuitive concepts of "skeleton" and prior effective sample size (PESS), and then describe how the similar approach can be used for model-assisted designs, including the BOIN and keyboard designs.

6.1.1 Incorporate prior information in CRM

Let π_j denote the true DLT probability of the jth dose of the J doses under investigation, $j = 1, \ldots, J$. We assume that the historical data or RWD have been summarized in the form of the prior estimate of π_j, denoted as q_j, $j = 1, \cdots, J$. The prior estimate q_j can be obtained by fitting a statistical model (e.g., a logistic model or nonparametric model discussed in Liu et al. (2015)) to the historical data, or provided by clinicians based on their clinical experience.

It is straightforward to incorporate the prior information into CRM through the following power model:

$$\pi_j = q_j^{\exp\{\alpha\}}, \text{ for } j = 1, \ldots, J, \tag{6.1}$$

DOI: 10.1201/9780429052781-6

where α is the unknown parameter. Recall that $q_1 < \cdots < q_J$ are known as the skeleton, see Section 2.5. Under the Bayesian paradigm, we assign α a normal prior $f(\alpha) = \mathrm{N}(0, \sigma^2)$, where σ^2 is a prespecified hyperparameter.

This model *a priori* centers the dose–toxicity curve $(\pi_1, \cdots, \pi_J)$ around the skeleton $(q_1, \cdots, q_J)$. The value of σ^2 controls the amount of information borrowed from historical data via the skeleton. A smaller value leads to stronger borrowing. If $\sigma^2 = 0$, the prior completely dominates the observed data with $\pi_j \equiv q_j$ regardless of the observed data.

To make the decision of dose escalation and de-escalation, CRM updates the posterior estimate of π_j based on the observed interim data $D = \{(n_j, y_j), j = 1, \ldots, J\}$:

$$\hat{\pi}_j = \int q_j^{\exp(\alpha)} \frac{L(D \mid \alpha) f(\alpha)}{\int L(D \mid \alpha) f(\alpha) d\alpha} d\alpha,$$

where $L(D \mid \alpha) = \prod_{j=1}^{J} \left\{ q_j^{\exp(\alpha)} \right\}^{y_j} \left\{ 1 - q_j^{\exp(\alpha)} \right\}^{n_j - y_j}$ is the likelihood function. Then, CRM assigns the next cohort of patients at the dose whose $\hat{\pi}_j$ is closest to the target toxicity rate ϕ. In practice, we typically impose safety rules, such as starting at the lowest dose level and no dose skipping during dose transition, see Section 2.5 for details.

To choose an appropriate σ^2, it is of great importance to quantify how much information is borrowed from historical data with the prior $f(\alpha) = \mathrm{N}(0, \sigma^2)$. Zhou et al. (2021a) proposed a simple and intuitive approach to formally quantify the information borrowed through the skeleton using the concept of PESS. The strategy is to approximate the prior distribution of π_j, induced by model (6.1) and $f(\alpha)$, with a beta distribution by matching the first and second moments. Given skeleton $(q_1, \cdots, q_J)$ and prior $f(\alpha)$, the mean μ_j and variance τ_j^2 of π_j is given by

$$\mu_j = \int \pi_j f(\pi_j) \mathrm{d}\pi_j, \qquad \tau_j^2 = \int \pi_j^2 f(\pi_j) \mathrm{d}\pi_j - \mu_j^2,$$

where $f(\pi_j)$ is the prior distribution of π_j induced by the prior distribution of $f(\alpha) = \mathrm{N}(0, \sigma^2)$ and the model (6.1), given by

$$f(\pi_j) = \frac{1}{\sqrt{2\pi}\sigma} \exp\left\{ -\frac{\left[\log\left(\frac{\log(\pi_j)}{\log(q_j)} \right) \right]^2}{2\sigma^2} \right\} \frac{1}{\pi_j \log(\pi_j)}.$$

Matching the first and second moments, $f(\pi_j)$ is approximated by $\mathrm{Beta}(a_j, b_j)$, where

$$a_j = \frac{\mu_j^2(1 - \mu_j)}{\tau_j^2} - \mu_j, \quad b_j = \frac{a_j(1 - \mu_j)}{\mu_j}. \tag{6.2}$$

As $y_j \sim \text{Binomial}(n_j, \pi_j)$, the posterior of π_j is $f(\pi_j | D) = \text{Beta}(a_j + y_j, b_j + n_j - y_j)$, therefore the PESS of $f(\pi_j)$ is $a_j + b_j$. An alternative more sophisticated method of calculating PESS is provided by Morita et al. (2008).

This reveals a property of CRM: once the prior $f(\alpha)$ is specified, PESS for each dose is automatically determined because π_j is a function of α. For example, given skeleton $(q_1, \cdots, q_5) = (0.10, 0.19, 0.30, 0.42, 0.54)$ and prior $f(\alpha) = N(0, 0.72)$, PESS is $(3, 3, 3, 3.1, 3.4)$ for the five doses. Since the dose–toxicity model of CRM usually has degrees of freedom smaller than the number of doses under investigation, CRM does not allow users to specify dose-specific prior information or PESS. However, in practice, we often have an unequal amount of prior information for different doses. For example, we often have more data at the doses that are below and around MTD from historical phase I trials. In this case, it is highly desirable to be able to specify a different PESS for each unique dose according to the historical data. Most model-based phase I designs, e.g., the designs based on escalation with overdose control (Babb et al., 1998) or the Bayesian logistic regression model (Neuenschwander et al., 2008), share this limitation.

6.2 BOIN with Informative Prior (iBOIN)

6.2.1 Trial design

We now discuss how to use the skeleton, coupled with PESS, to incorporate prior information into the BOIN design. Let $\hat{\pi}_j = y_j/n_j$ denote the observed DLT rate at the current dose, and λ_{ej} and λ_{dj} denote the dose escalation and de-escalation boundaries, respectively. As shown in Section 3.4.2 and using the same notations, BOIN's optimal λ_{ej} and λ_{dj} for the observed DLT rate $\hat{\pi}_j$ are given by

$$\lambda_{ej} = \max\left\{0, \frac{\log\left(\frac{1-\phi_1}{1-\phi}\right) + n_j^{-1}\log\left(\frac{\omega_{1j}}{\omega_{0j}}\right)}{\log\left(\frac{\phi(1-\phi_1)}{\phi_1(1-\phi)}\right)}\right\}, \quad (6.3)$$

$$\lambda_{dj} = \min\left\{1, \frac{\log\left(\frac{1-\phi}{1-\phi_2}\right) + n_j^{-1}\log\left(\frac{\omega_{0j}}{\omega_{2j}}\right)}{\log\left(\frac{\phi_2(1-\phi)}{\phi(1-\phi_2)}\right)}\right\}, \quad (6.4)$$

with the constraint $\lambda_{ej} \leq \lambda_{dj}$. When the non-informative prior $\omega_{0j} = \omega_{1j} = \omega_{2j} = 1/3$ is used, which is recommended for most trials, the optimal escalation and de-escalation boundaries become

$$\lambda_e \equiv \lambda_{ej} = \frac{\log\left(\frac{1-\phi_1}{1-\phi}\right)}{\log\left(\frac{\phi(1-\phi_1)}{\phi_1(1-\phi)}\right)}, \qquad \lambda_d \equiv \lambda_{dj} = \frac{\log\left(\frac{1-\phi}{1-\phi_2}\right)}{\log\left(\frac{\phi_2(1-\phi)}{\phi(1-\phi_2)}\right)},$$

which are independent of j and n_j. These are boundaries used in the standard BOIN, see Section 3.4. It can be shown that $\lambda_e \leq \lambda_d$.

When historical data (or reliable prior information) are available, they can be incorporated into BOIN based on the following steps:

1. Estimate skeleton $(q_1, \cdots, q_J)$ from the historical data and elicit corresponding PESS $(n_{01}, \cdots, n_{0J})$, where n_{0j} is the desirable PESS for dose level j, $j = 1, \cdots, J$.

2. Determine the informative prior for H_k, i.e., ω_{kj}, $j = 0, 1, 2$, as

$$\omega_{kj} = \sum_{x=0}^{n_{0j}} \frac{\varphi_k^x (1-\varphi_k)^{n_{0j}-x}}{\sum_{k'=0}^{2} \varphi_{k'}^x (1-\varphi_{k'})^{n_{0j}-x}} \binom{n_{0j}}{x} q_j^x (1-q_j)^{n_{0j}-x}, \quad (6.5)$$

   where $\varphi_0 = \phi$, $\varphi_1 = \phi_1$, and $\varphi_2 = \phi_2$.

3. Calculate $(\lambda_{ej}, \lambda_{dj})$ using formulas (6.3) and (6.4). In the rare case that formulas (6.3) and (6.4) result in a solution $\lambda_{ej} > \lambda_{dj}$, then determine the optimal $(\lambda_{ej}, \lambda_{dj})$ using a numerical search to minimize the decision error.

The derivation of ω_{kj} in Step 2 is as below. For dose j, the predictive probability of H_{kj} based on the prior DLT rate q_j and PESS n_j can be expressed as

$$\begin{aligned}
\omega_{kj} &= \Pr(H_{kj} \mid n_{0j}, q_j) \\
&= \sum_{x=0}^{n_{0j}} \Pr(H_{kj} \mid x) \Pr(x \mid n_{0j}, q_j) \\
&= \sum_{x=0}^{n_{0j}} \frac{\Pr(x \mid H_{kj}) \Pr(H_k)}{\sum_{k'=0}^{2} \Pr(x \mid H_{k'}) \Pr(H_{k'})} \Pr(x \mid n_{0j}, q_j) \\
&= \sum_{x=0}^{n_{0j}} \frac{\varphi_k^x (1-\varphi_k)^{n_{0j}-x}}{\sum_{k'=0}^{2} \varphi_{k'}^x (1-\varphi_{k'})^{n_{0j}-x}} \binom{n_{0j}}{x} q_j^x (1-q_j)^{n_{0j}-x}.
\end{aligned}$$

The last equality follows by assuming $\Pr(H_0) = \Pr(H_1) = \Pr(H_2) = 1/3$.

In Step 3, under certain extreme specifications of the prior information (e.g., extremely informative prior), it is possible that the $\lambda_{ej} > \lambda_{dj}$ based on formulas (6.3) and (6.4). In this case, the escalation and de-escalation boundaries $(\lambda_{ej}, \lambda_{dj})$ should be re-determined using a numerical search to minimize the decision error, given by equation (3.3) in Section 3.4, under the constraint of $\lambda_{ej} \leq \lambda_{dj}$. This numerical search is fast and straightforward given that both the possible numbers of patients treated and DLTs at a dose are limited (e.g., < 20).

We refer to the resulting design (with informative prior) based on $(\lambda_{ej}, \lambda_{dj})$ as iBOIN. Because of the incorporation of the informative prior information, the escalation and de-escalation boundaries λ_{ej} and λ_{dj} of iBOIN depend on the dose level j, as well as n_j. When $(n_{01}, \cdots, n_{0J}) = 0$, the value of ω_{kj} becomes $1/3$ using the above equation, as a result iBOIN reduces to the standard BOIN design in this case.

Figure 6.1 contrasts the boundaries under a non-informative prior and those under an informative prior for a trial with five doses and an elicited skeleton (0.10, 0.19, 0.30, 0.42, 0.54) when the target DLT probability is 0.3 and PESS is 3. For example, because the prior information says that the lowest dose is below the true MTD (with the prior DLT probability of 0.10), its escalation boundary λ_{ej} is higher than that of the non-informative prior to encourage dose escalation. On the contrary, because the prior information says the highest dose is above MTD (with the prior DLT probability of 0.54), its de-escalation boundary λ_{dj} is lower than that of the non-informative prior to encourage dose de-escalation.

Compared to CRM/EWOC/BLRM, iBOIN is more flexible and allows users to accurately incorporate prior information by specifying PESS for each dose. For example, given a phase I trial with five doses, if historical data provide more information on the first two doses than the last two doses and most information on dose level 3, we could specify the five doses' PESS as

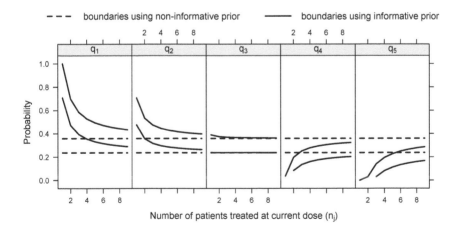

FIGURE 6.1: Dose escalation and de-escalation boundaries under the non-informative prior and an informative prior with the skeleton $(q_1, q_2, q_3, q_4, q_5) = (0.10, 0.19, 0.30, 0.42, 0.54)$, PESS $= 3$, and $\phi = 0.3$.

(3, 3, 6, 1, 1) to reflect that. As described previously, this is not possible under CRM.

The other advantage of iBOIN, as a model-assisted design, is that the dose escalation and de-escalation rule can be pre-tabulated and included in the trial protocol. Table 6.1 shows the decision table of iBOIN with the skeleton (0.10, 0.19, 0.30, 0.42, 0.54) and PESS $n_{01} = \cdots = n_{05} = 3$. This decision table is equivalent to the rule based on λ_{ej} and λ_{dj}, but easier to use in practice. Users only need to identify the row corresponding to the current dose level, and then they can use the boundaries listed in that row to easily make the decision of dose escalation and de-escalation.

In summary, the dose-finding rules of the iBOIN design can be described as follows:

1. Patients in the first cohort are treated at the lowest dose d_1, or the physician-specified dose.

2. Given data (n_j, y_j) observed at the current dose level j, make the decision of escalation/de-escalation according to the iBOIN decision table (e.g., Table 6.1) for treating the next cohort of patients.

3. Repeat Step 2 until the prespecified maximum sample size is reached, and then select MTD as the dose whose isotonic estimate of π_j is closest to ϕ.

For the purpose of overdose control, following BOIN, the iBOIN design imposes a dose elimination rule: if $\Pr(\pi_j > \phi \,|\, n_j, y_j) > 0.95$ and $n_j \geq 3$, dose

TABLE 6.1: iBOIN decision boundaries up to 30 patients with a cohort size of three, given the skeleton $(q_1, \cdots, q_5) = (0.10, 0.19, 0.30, 0.42, 0.54)$ and PESS $n_{01} = \cdots = n_{05} = 3$. The target DLT probability $\phi = 0.3$.

Dose		No. of patients at current dose									
level	Action*	3	6	9	12	15	18	21	24	27	30
1	Escalate if no. of DLT $\leq$	1	1	2	3	4	4	5	6	6	7
	De-escalate if no. of DLT $\geq$	2	3	4	5	7	8	9	10	11	12
2	Escalate if no. of DLT $\leq$	0	1	2	3	3	4	5	5	6	7
	De-escalate if no. of DLT $\geq$	2	3	4	5	6	7	8	9	11	12
3	Escalate if no. of DLT $\leq$	0	1	2	2	3	4	4	5	6	7
	De-escalate if no. of DLT $\geq$	2	3	4	5	6	7	8	9	10	11
4	Escalate if no. of DLT$\leq$	0	1	1	2	3	3	4	5	6	6
	De-escalate if no. of DLT $\geq$	1	2	3	4	6	7	8	9	10	11
5	Escalate if no. of DLT $\leq$	0	0	1	2	2	3	4	5	5	6
	De-escalate if no. of DLT $\geq$	1	2	3	4	5	6	7	8	10	11

*When neither "Escalate" nor "De-escalate" is triggered, stay at the current dose for treating the next cohort of patients.

level j and higher are eliminated from the trial, where $\Pr(\pi_j > \phi \mid n_j, y_j)$ is evaluated based on the beta-binomial model with the Unif$(0,1)$ prior. As the objective of the dose elimination rule is to protect patients from excessively toxic doses, it is sensible to use the uniform prior to evaluate this rule to avoid potential bias due to misspecification of the prior. As discussed in Section 3.4.5, if desirable, a vague prior such as Beta$(0.02, 0.08)$ with the PESS $= 0.1$ can be also used to obtain similar operating characteristics on dose elimination by slightly adjusting the overdose control probability cutoff 0.95 (e.g., to 0.93). The trial is terminated if the lowest dose level is eliminated.

6.2.2 Practical guidance

Choosing PESS

As the skeleton is typically estimated based on the history data, when applying iBOIN, the key question is: "How to specify PESS?" PESS should be chosen to reflect the appropriate amount of prior information to be incorporated, which depends on the reliability of the prior information and varies from trial to trial. When there is strong evidence that the prior is most likely to be specified correctly, it is appropriate to use a large PESS to borrow more information; when there is a great amount of uncertainty regarding whether the prior is most likely correctly specified, we may use a small PESS to avoid bias.

In practice, there is often sizable uncertainty on the reliability of the prior information. Thus, PESS should be chosen carefully to achieve an appropriate balance between design performance and robustness. Using a large PESS improves the design performance (i.e., the accuracy to identify MTD) when the prior is correctly specified, but may lead to a substantial loss of performance when the prior is grossly misspecified. Based on numerical studies, Zhou et al. (2021a) recommended PESS $\in [1/3(N/J), 1/2(N/J)]$ as the default value that improves trial performance while maintaining reasonable robustness. For example, when $J = 5$ and $N = 30$, the recommended value for PESS is $n_{0j} = 2$ or 3 (i.e., across five doses, the total PESS is 10 or 15). The value of n_{0j} can be further calibrated by simulation using the software described in Section 6.5.

Robust prior

In general, iBOIN is robust to the misspecification of prior information (i.e., prior-data conflict), especially when PESS is chosen as described above. However, when the informative prior is grossly misspecified (e.g., the prior estimate of MTD and the true MTD differ by three dose levels), it may compromise the accuracy of identifying MTD. To further strengthen the robustness of iBOIN, Zhou et al. (2021a) proposed a robust prior, which is easy to implement and yields superior operating characteristics.

Given the elicited skeleton $(q_1, \cdots, q_J)$ with dose level j^* as the prior estimate of MTD (i.e., $q_{j^*} = \phi$), the robust prior is the same as the prior described above when $j^* < J/2$, but modifies PESS to $(n_{01}, \cdots, n_{0j^*}, 0, \cdots, 0)$ when $j^* \geq J/2$. In other words, when prior MTD $j^* \geq J/2$, the robust prior uses informative prior information for the dose up to the prior MTD estimate, and after that it uses the non-informative prior.

The rationale of this robust prior is that the dose finding is a sequential process of allocating patients from low doses to high doses. Thus, by the time that dose finding reaches high doses, there is an extremely limited sample size remaining to override the prior if it is misspecified. The robust prior modifies the prior of high doses to be non-informative to facilitate overriding the prior when the data conflict with the prior, thus alleviating the impact of prior misspecification. This prior is particularly useful when there is a great amount of uncertainty regarding the prior information. Zhou et al. (2021a) also considered another robust prior, which is a mixture of informative and non-informative priors. This mixture robust prior is conceptually appealing, but more complicated and does not perform as well as the simple robust prior described above.

6.3   iKeyboard design

It is straightforward to incorporate the prior information into the keyboard design using the skeleton and PESS. As described in Section 3.3, the keyboard design assumes a beta-binomial model,

$$
\begin{aligned}
y_j \mid n_j, \pi_j &\sim \text{Binomial}(n_j, \pi_j), \\
\pi_j &\sim \text{Beta}(a_j, b_j),
\end{aligned}
\tag{6.6}
$$

where a_j and b_j are hyperparameters. Let $D_j = (n_j, y_j)$, the posterior distribution of π_j arises as

$$
\pi_j \mid D_j \sim \text{Beta}(y_j + a_j, n_j - y_j + b_j), \text{ for } j = 1, \ldots, J.
\tag{6.7}
$$

The keyboard design divides the toxicity probability line $(0, 1)$ into a series of equal width intervals, and makes the decision of dose escalation and de-escalation by comparing the location of the strongest key $\mathcal{I}_{\max}$, defined as the interval with the highest posterior probability, with the target key $\mathcal{I}^* = (\delta_1, \delta_2)$ (i.e., the pre-specified target dosing interval). If $\mathcal{I}_{\max}$ is located on the left side of $\mathcal{I}^*$ (denoted as $\mathcal{I}_{\max} \prec \mathcal{I}_{\text{target}}$), escalate the dose; if $\mathcal{I}_{\max}$ is located on the right side of $\mathcal{I}^*$ (denoted as $\mathcal{I}_{\max} \succ \mathcal{I}_{\text{target}}$), de-escalate the dose; if $\mathcal{I}_{\max}$ is $\mathcal{I}^*$ (denoted $\mathcal{I}_{\max} \equiv \mathcal{I}_{\text{target}}$), retain the current dose. See Section 3.3 for more details.

In the beta-binomial model (6.6), $a_j + b_j$ can be interpreted as PESS. Thus, the prior information can be incorporated into the keyboard design as follows:

1. Estimate skeleton $(q_1, \cdots, q_J)$ from the historical data and elicit corresponding PESS $(n_{01}, \cdots, n_{0J})$, where n_{0j} is the desirable PESS for dose level j, $j = 1, \cdots, J$.

2. Determine hyperparameters a_j and b_j in the beta prior (6.6) as follows:

$$a_j = n_{0j}q_j; \qquad b_j = n_{0j}(1 - q_j), \qquad j = 1, \cdots, J.$$

3. Make dose escalation and de-escalation based on the resulting posterior given by equation (6.7).

We refer to the keyboard design with an informative prior as the iKeyboard design. Given a fixed maximum sample size, all possible outcomes $D_j = (n_j, y_j)$ can be enumerated. For each possible outcome, the posterior distribution $f(\pi_j | D_j)$ can be calculated. Therefore, the dose escalation/de-escalation rule of iKeyboard can be tabulated, similar to iBOIN. The above approach is directly applicable to the mTPI design for incorporating prior information.

6.4 Operating characteristics

In this section, we briefly describe a simulation study to evaluate the operating characteristics of the iCRM (i.e., CRM with informative prior), iBOIN, and iKeyboard designs. A comprehensive simulation study (including comparison using random scenarios) is provided by Zhou et al. (2021a). Consider $J = 5$ doses and the target DLT probability $\phi = 0.3$. The maximum sample size is $N = 30$ with a cohort size of three. For iBOIN and iKeyboard, we set PESS $n_{0j} = 3$ for $j = 1, \cdots, 5$; and for iCRM, the prior is chosen such that PESS at the prior MTD is three. All the designs use the same skeletons (i.e., the prior DLT probabilities), which are provided in Table 6.2. The counterparts of the designs with a non-informative prior (denoted as CRM, BOIN, and keyboard) are included for comparison.

Table 6.3 shows the results, including (1) percentage of correct selection (PCS), defined as the percentage of simulated trials in which MTD is correctly identified; (2) percentage of patients treated at MTD; (3) percentage of patients treated above MTD; (4) risk of overdosing, defined as the percentage of simulated trials that assigned 50% or more patients to doses above MTD; and (5) risk of poor allocation, defined as the percentage of simulated trials that assigned fewer than six patients to MTD. As noted by Zhou et al. (2018b), metrics (4) to (5) measure the reliability of the design, i.e., the likelihood of a design demonstrating extremely problematic behaviors (e.g., treating 50% or more patients at toxic doses, or fewer than six patients at MTD), which are of great practical importance. Note that the percentage of patients overdosed (i.e., metric (3)) does not completely capture the risk of overdosing (i.e., metric (4)). Two designs can have a similar percentage of patients overdosed, but rather different risks of overdosing 50% of the patients.

TABLE 6.2: Four dose–toxicity scenarios with target DLT probability $\phi =$ 0.30. The prior MTDs are correctly specified in Scenarios 1 and 2 and mis-specified in Scenarios 3 and 4.

	Dose level				
	1	2	3	4	5
			Scenario 1		
True Pr(DLT)	*0.30*	0.42	0.50	0.60	0.65
Prior Pr(DLT)	*0.30*	0.42	0.54	0.64	0.73
			Scenario 2		
True Pr(DLT)	0.15	*0.27*	0.40	0.50	0.65
Prior Pr(DLT)	0.19	*0.30*	0.42	0.54	0.64
			Scenario 3		
True Pr(DLT)	0.09	0.12	0.15	*0.30*	0.45
Prior Pr(DLT)	0.01	0.04	0.10	0.19	*0.30*
			Scenario 4		
True Pr(DLT)	0.08	0.15	*0.31*	0.45	0.55
Prior Pr(DLT)	0.19	*0.30*	0.42	0.54	0.64

The simulation results show that (i) incorporating prior information improves the accuracy of identifying MTD for all designs; (ii) iBOIN and iCRM have similar performance in identifying MTD and allocating patients to MTD, but iBOIN has a lower risk of overdosing patients and poor allocation; (iii) Compared to iBOIN, iKeyboard has a larger variation in identifying MTD (e.g., performs best in scenario 3, but worst in scenario 4), and a higher risk of overdosing patients and poor allocation.

6.5   Software and case study

The web application for iBOIN is available as a module of the "BOIN Suite" at http://www.trialdesign.org. Figure 6.2 shows the launchpad of BOIN Suite.

Non-small Cell Lung Cancer (NSCLC) Cancer Trial   The objective of this phase I trial (ClinicalTrials.gov Identifier: NCT04479306) is to determine MTD and RP2D of alisertib in combination with osimertinib in treating patients with EGFR mutated stage IIIB or IV NSCLC. Toxicity will be evaluated according to the National Cancer Institute Common Terminology Criteria for Adverse Events (NCI CTCAE) version 5.0. DLT will be assessed during the first 28-day cycle of combination therapy. The maximum sample size is 12 and patients are treated in a cohort size of three. Three doses of alisertib will be investigated in combination with osimertinib, and the target DLT probability is $\phi = 30\%$.

We use this trial as an example to illustrate how to use iBOIN to leverage the prior information to improve the efficacy of phase I trials. After selecting and

TABLE 6.3: Operating characteristics of iCRM, iBOIN, and iKeyboard in comparison with their counterparts with non-informative priors. PCS is the percentage of correction selection, ROD is the risk of overdosing, and RPA is the risk of poor allocation.

Design	PCS	% Pts at MTD	% Pts >MTD	ROD	RPA
		Scenario 1			
CRM	54.8	59.9	27.8	23.2	12.2
iCRM	63.1	65.2	24.9	19.4	9.8
BOIN	59.2	59.6	29.0	23.6	10.2
iBOIN	64.2	66.2	22.4	12.8	4.5
Keyboard	59.2	59.3	29.3	23.6	10.2
iKeyboard	64.2	50.7	39.6	34.2	17.8
		Scenario 2			
CRM	51.6	36.1	7.7	29.5	25.2
iCRM	53.3	42.4	5.6	23.7	16.8
BOIN	50.6	41.1	6.0	23.0	17.1
iBOIN	57.8	47.6	3.7	10.4	8.6
Keyboard	50.2	41.1	6.0	23.0	16.7
iKeyboard	59.6	37.8	6.6	35.1	23.6
		Scenario 3			
CRM	50.7	29.9	14.6	9.3	32.9
iCRM	57.3	33.8	17.0	13.0	28.2
BOIN	51.5	28.6	13.1	1.2	24.6
iBOIN	58.6	35.5	18.4	3.8	11.8
Keyboard	52.1	8.6	13.1	1.2	24.6
iKeyboard	59.5	9.2	27.9	11.2	17.9
		Scenario 4			
CRM	58.0	38.1	21.7	17.3	21.8
iCRM	59.8	38.0	18.3	13.4	21.3
BOIN	52.3	35.6	17.0	7.9	19.2
iBOIN	61.6	33.0	10.9	2.2	14.8
Keyboard	52.4	35.7	17.1	7.9	18.9
iKeyboard	56.4	45.4	15.3	3.6	6.8

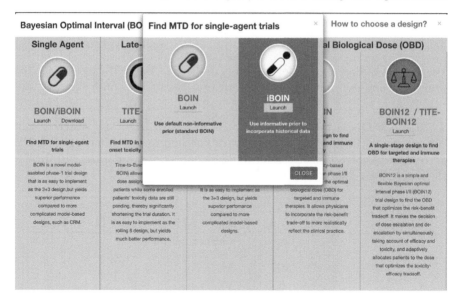

FIGURE 6.2: The launchpad of web application BOIN Suite available at http://www.trialdesign.org.

launching the "iBOIN" module from the BOIN Suite launchpad, we design the trial by taking the following three steps:

Step 1: Enter trial parameters

Doses, Sample Size, and Target   As shown in Figure 6.3, the number of doses under investigation is three, starting dose level is one. The starting dose level can be adjusted according to prior information, e.g., one level below the prior MTD estimated using historical data. For example, in some trials, the appropriate starting dose level may be two or even higher.

The cohort size for the trial is three and the number of cohorts is four, with a total sample size of 12. As a rule of thumb, we recommend the maximum sample size $N = 6 \times J$ (i.e., the maximum sample size of the 3+3 design) as the total sample size, where J is the number of doses. This trial uses a smaller sample size due to the limited accrual and the intention to shorten the trial duration. In this case, leveraging available prior information provides a useful approach to compensate the small sample size.

To reduce the sample size, it is often useful to use the "convergence" stopping rule: stop the trial early when m patients have been assigned to a dose and the decision is to stay at that dose. This stopping criterion suggests that the dose finding approximately converges to MTD, thus the trial can be stopped. We recommend $m = 9$ or larger. In this trial, $m = 9$ is used. Because of the early-stopping rule, the actual sample size used in the trial is often smaller than N. The saving in the sample size depends on the true

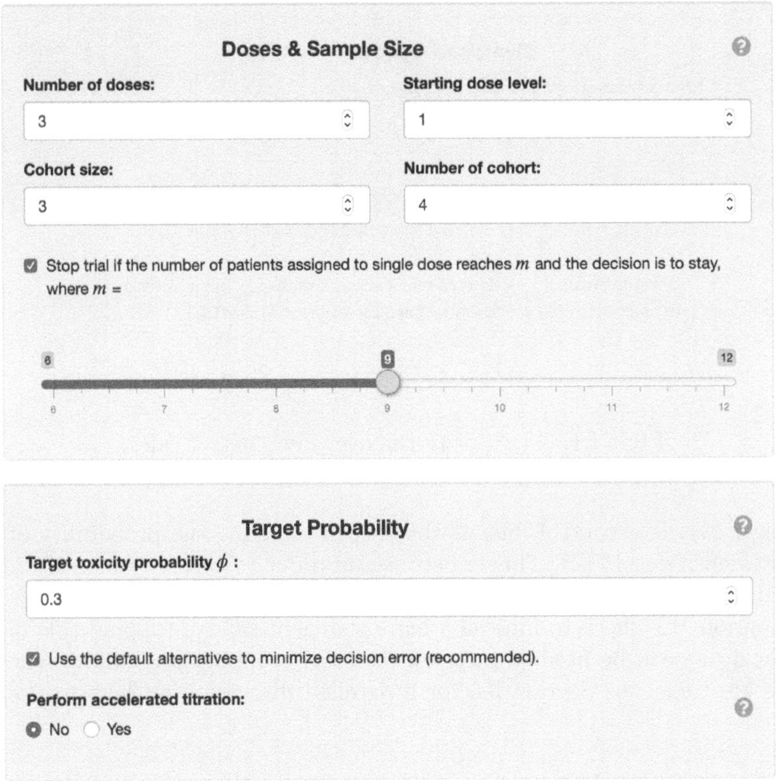

FIGURE 6.3: Specify doses, sample size, target, convergence stopping rule, and accelerated titration.

dose–toxicity scenario and can be evaluated using simulation. In general, a larger sample size saving is expected when the starting dose is close to the target.

The target DLT probability is $\phi = 0.3$, which should be elicited from clinicians. If a specific safety requirement is desirable, the values for ϕ, ϕ_1, and ϕ_2 can be further adjusted. In addition, as described previously, the accelerated titration option is another tool to reduce the sample size, see Section 3.6 for details.

Overdose Control   This panel (see Figure 6.4) specifies the overdose control rule, described in Section 5.4. That is, if $\Pr(\pi_j > \phi \,|\, D_j) > 0.95$ and $n_j \geq 3$, dose level j and higher are eliminated from the trial, and the trial is terminated if the lowest dose level is eliminated. When the trial is terminated due to toxicity, no dose should be selected as MTD. In general, we recommend to use the default probability cutoff 0.95. A smaller value, e.g., 0.9, results in

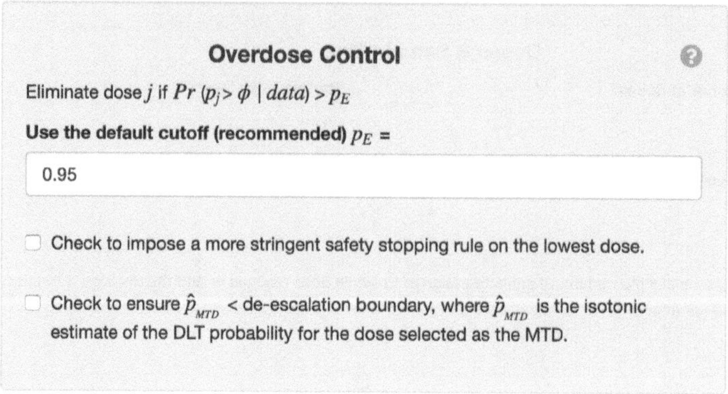

FIGURE 6.4: Specify the overdose control rules.

stronger overdose control, but at the cost of reducing the probability of correctly identifying MTD. This is because, in order to correctly identify MTD, it is imperative to explore the doses sufficiently to learn their toxicity profile. The option "☐ Check to impose a more stringent safety stopping rule on the lowest dose" can be used to increase the early stopping probability when all doses are toxic; see Section 3.6 for a detailed discussion on how to use this option.

Prior and PESS   This panel (see Figure 6.5) specifies the prior estimate of toxicity probability at each dose (i.e., the skeleton estimated based on the historical data) and PESS. Based on the historical data on alisertib and osimertinib, the combination is expected to be safe with little overlapped toxicities,

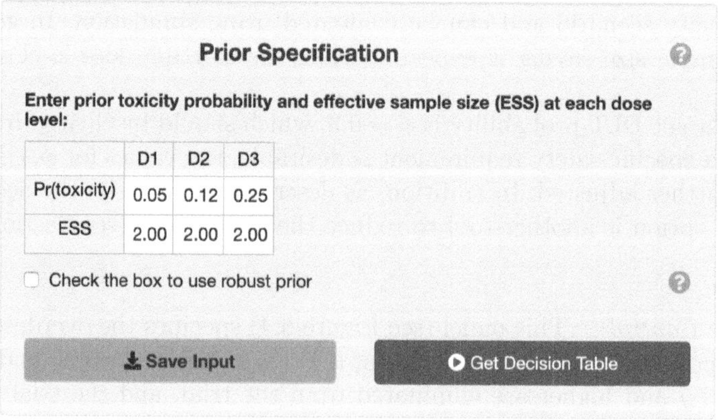

FIGURE 6.5: Specify the skeleton and PESS.

and thus the skeleton is set as (0.05, 0.12, 0.25) for the three doses. Following the guidance described in Section 6.2.2, PESS is set as two for the three doses, leading to a prior being informative without dominating the trial data. For many trials, it is desirable to use the robust prior to obtain extra robustness. This can be done by checking the corresponding activation box. For this trial, the robust prior is identical to the informative prior, as the prior estimate of MTD is the last dose.

After completing the specification of trial parameters, the decision table will be generated by clicking the "Get Decision Table" button. The decision table will be automatically included in the protocol template in Step 3, but can also be saved as a separate csv, Excel or pdf file in this step if needed.

The resulting dose escalation and de-escalation boundaries are shown in Table 6.4. As the prior information suggests that dose level one is far below MTD, the escalation boundary of dose level one (e.g., escalate if $\leq 1/3$ DLT) is higher than that of dose levels two and three (e.g., escalate if $\leq 0/3$ DLT), making it easier to escalate at dose level 1. When more patients are treated, the trial data start to dominate the prior, and thus the escalation and de-escalation boundaries become the same for three doses when six or nine patients are treated at a dose. These boundaries eventually shrink to the standard BOIN boundaries when the sample size is sufficiently large.

Step 2: Run simulation

Operating Characteristics   This step generates the operating characteristics of the design through simulation, see Figure 6.6. The scenarios used for sim-

TABLE 6.4: iBOIN decision boundaries for the lung cancer trial, given the skeleton $(q_1, \cdots, q_5) = (0.05, 0.12, 0.25)$ and PESS $n_{01} = n_{02} = n_{03} = 2$. The target DLT probability $\phi = 0.3$.

Dose		No. of patients at current dose		
level	Action*	3	6	9
	Escalate if No. of DLT $\leq$	1	1	2
1	De-escalate if No. of DLT $\geq$	2	3	4
	Eliminate if No. of DLT $\geq$	3	4	5
	Escalate if No. of DLT $\leq$	0	1	2
2	De-escalate if No. of DLT $\geq$	2	3	4
	Eliminate if No. of DLT $\geq$	3	4	5
	Escalate if No. of DLT $\leq$	0	1	2
3	De-escalate if No. of DLT $\geq$	2	3	4
	Eliminate if No. of DLT $\geq$	3	4	5

*When neither "Escalate" nor "De-escalate" is triggered, stay at the current dose for treating the next cohort of patients.

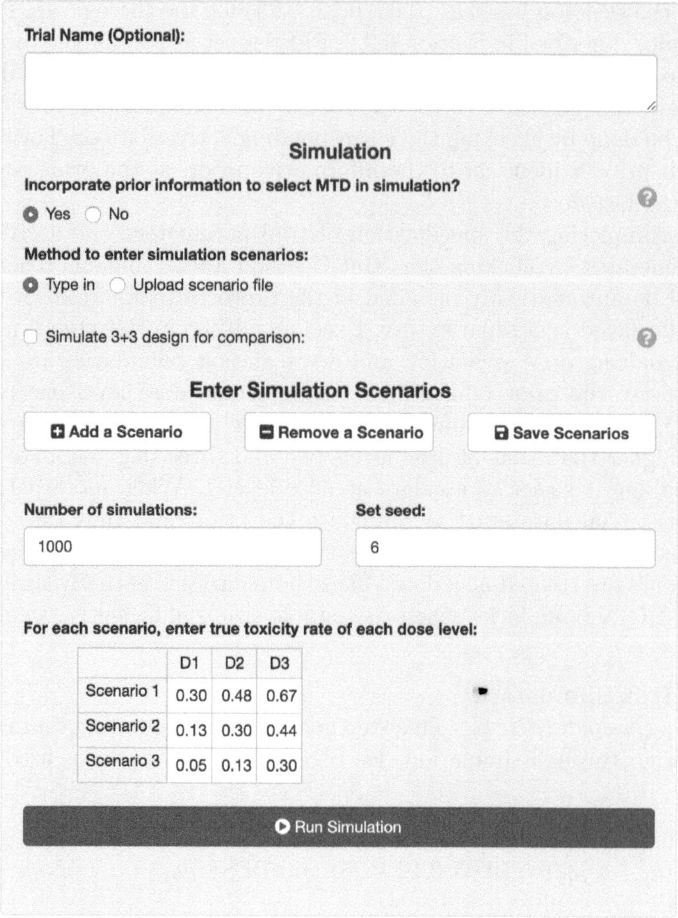

Trial Name (Optional):

Simulation

Incorporate prior information to select MTD in simulation?

● Yes ○ No

Method to enter simulation scenarios:

● Type in ○ Upload scenario file

☐ Simulate 3+3 design for comparison:

Enter Simulation Scenarios

➕ Add a Scenario	➖ Remove a Scenario	🖫 Save Scenarios

Number of simulations: **Set seed:**

1000   6

For each scenario, enter true toxicity rate of each dose level:

	D1	D2	D3
Scenario 1	0.30	0.48	0.67
Scenario 2	0.13	0.30	0.44
Scenario 3	0.05	0.13	0.30

▶ Run Simulation

FIGURE 6.6: Simulate operating characteristics of the design.

ulation should cover various possible clinical scenarios, e.g., MTD is located at different dose levels. To facilitate the generation of the operating characteristics of the design, the software automatically provides a set of randomly generated scenarios with various MTD locations, which are often adequate for most trials. Depending on the application, users can add or remove scenarios. The software also provides an option to include the 3+3 design as a comparator to facilitate communication with clinicians who are more familiar with the conventional design. Table 6.5 shows the operating characteristics of iBOIN compared with those of BOIN using the non-informative prior. When the MTD prior estimate is correctly specified, iBOIN improves the percentage of correct selection from 61.4% to 72.5%. The simulation results will be

TABLE 6.5: Operating characteristics of the iBOIN design for the lung cancer trial in comparison with the BOIN with the non-informative prior. The dose in bold is MTD.

Design		Dose level		
		1	2	3
	Pr(DLT)	0.05	0.13	**0.3**
BOIN	Selection (%)	1.5	37.1	**61.4**
	% of Patients	30.6	38.3	**31.0**
iBOIN	Selection (%)	1.4	26.1	**72.5**
	% of Patients	27.2	39.2	**33.6**
	Pr(DLT)	0.13	**0.3**	0.44
BOIN	Selection (%)	23.8	**53.3**	22.5
	% of Patients	43.1	**43.1**	13.8
iBOIN	Selection (%)	15.0	**53.5**	30.9
	% of Patients	36.3	**47.1**	16.6
	Pr(DLT)	**0.3**	0.48	0.67
BOIN	Selection (%)	**60.2**	23.4	2.9
	% of Patients	**68.5**	28.7	2.8
iBOIN	Selection (%)	**54.0**	31.3	3.9
	% of Patients	**56.7**	38.8	4.5

automatically included as a table in the protocol template in the next step, but can also be saved as a separate csv or Excel file if needed.

Step 3: Generate protocol template

Protocol Preparation   The iBOIN software generates sample texts and a protocol template to facilitate the protocol write-up. The protocol template can be downloaded in various formats (see Figure 6.7). Use of this module requires the completion of Steps 1 and 2. Once the protocol is approved by regulatory

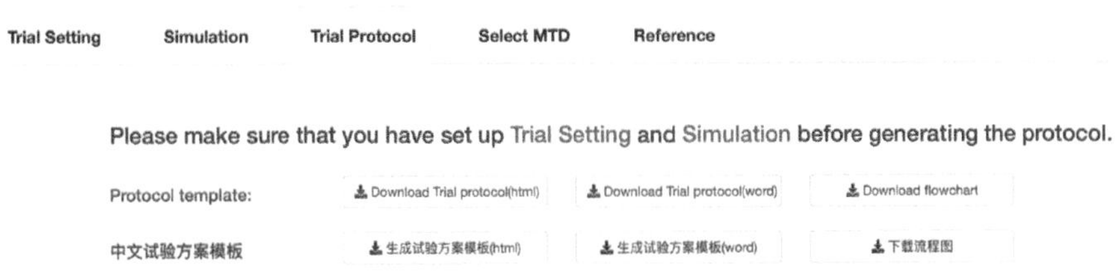

Trial Setting   Simulation   Trial Protocol   Select MTD   Reference

Please make sure that you have set up Trial Setting and Simulation **before generating the protocol.**

Protocol template: ⬇ Download Trial protocol(html)   ⬇ Download Trial protocol(word)   ⬇ Download flowchart

中文试验方案模板   ⬇ 生成试验方案模板(html)   ⬇ 生成试验方案模板(word)   ⬇ 下载流程图

FIGURE 6.7: Download protocol templates.

bodies (e.g., *Institutional Review Board*), we follow the design decision table to conduct the trial and make adaptive decisions (e.g., dose escalation/stay/de-escalation). At the completion of the trial, the "Select MTD" function can be used to determine MTD based on the observed trial data.

7

Multiple Toxicity Grades

7.1 Multiple toxicity grades

The landscape of oncology drug development has recently changed with the emergence of molecularly targeted agents and immunotherapies. These new therapeutic agents appear more likely to induce multiple low- or moderate-grade toxicities rather than the dose-limiting toxicity (DLT) (Brahmer et al., 2010; Le Tourneau et al., 2010; Penel et al., 2011). To accurately account for the side effects of the agents, it is important to incorporate the grade of toxicity into dose finding and decision making. The widely used National Cancer Institute Common Terminology Criteria for Adverse Events (NCI CTCAE) scores toxicity as a five-grade ordinal variable. Its general guidelines use grade 0 for no toxicity, grade 1 for mild toxicity, grade 2 for moderate toxicity, grade 3 for severe toxicity, grade 4 for life-threatening toxicity, and grade 5 for toxicity-related death.

The designs introduced so far, including the 3+3, continual reassessment method (CRM), and Bayesian optimal interval (BOIN) designs, all assume a single binary outcome: DLT (e.g., grade 3 or higher toxicity) or not. This dichotomization disregards low-grade adverse events, and it thus may be insufficient to quantify the safety of the targeted agents and immunotherapies that cause a high frequency of low- or moderate-grade toxicities with little DLT. The aggregate effect of multiple lower- or moderate-grade toxicity events, however, often leads to dose interruption, reduction, and discontinuation, which reduces the efficacy of the drug in the real-world setting.

Three approaches have been proposed to incorporate toxicity grades into dose finding, including total toxicity burden, toxicity score, and multiple toxicities.

(1)Total toxicity burden

One general approach is to assign a severity weight to each grade and type of toxicity event, and then combine the weights as a composite score. Bekele and Thall (2004) proposed the total toxicity burden (TTB) as the arithmetic sum of different grades and types of toxicity, weighted by the severity weights elicited from clinicians to reflect the importance among the toxicities that the clinicians had identified. For example, Table 7.1 shows the severity weights for the soft tissue sarcoma phase I trial described by Bekele and Thall (2004). If a patient experienced a grade 3 myelosuppression without fever, grade 3

DOI: 10.1201/9780429052781-7

TABLE 7.1: Toxicities and severity weights in the sarcoma trial, reproduced from Bekele and Thall (2004).

	Type of toxicity	Grade	Severity weight
1	Myelosuppression without fever	3	1.0
		4	1.5
	Myelosuppression with fever	3	5.0
		4	6.0
2	Dermatitis	3	2.5
		4	6.0
3	Liver	2	2.0
		3	3.0
		4	6.0
4	Nausea/vomiting	3	1.5
		4	2.0
5	Fatigue	3	0.5
		4	1.0

dermatitis, and grade 3 nausea/vomiting, the TTB for that patient is then calculated as TTB $= 1 + 2.5 + 1.5 = 5$. The maximum tolerated dose (MTD) is defined as the dose level where the true TTB value is closest to the target TTB value specified by the oncologists. To identify the MTD, Bekele and Thall (2004) built a joint model for the toxicities under the Bayesian framework. At each decision-making time, the dose that has the posterior expected TTB closest to the target TTB is selected as the next level.

Similarly, Chen et al. (2010) proposed to map toxicity grades into a quasi-continuous or continuous endpoint using the normalized equivalent toxicity score (NETS). Lee et al. (2009) proposed the toxicity burden score (TBS) to summarize toxicity using a weighted sum, where the severity weights were estimated via regression based on historical data. Ezzalfani et al. (2013) proposed another flexible toxicity endpoint, called the total toxicity profile (TTP), which is computed as the Euclidean norm of the severity weights. Although these composite scores (TTB, NETS, TBS, and TTP) are not strictly continuous, in practice, after appropriate transformation, they can be approximately regarded as a normally distributed endpoint.

(2) Toxicity score

Several methods have been proposed to "convert" toxicity grades to numeric scores that reflect their relative severity in the unit of DLT. For example, we may define a grade 3 toxicity as equivalent to a DLT, a grade 2 toxicity as equivalent to a 0.5 DLT, and a grade 4 toxicity as equivalent to 2 DLTs. This is the "equivalent toxicity score" (ETS) approach proposed by Yuan et al. (2007). Although the (normalized) ETS is a fractional event and does not really follow a Bernoulli distribution, Yuan et al. (2007) showed that it can be simply treated as a binary endpoint (more precisely, a quasi-binary end-point) and modeled using the quasi-Bernoulli likelihood. Yuan et al. (2007) implemented this quasi-Bernoulli likelihood in the CRM framework, leading to a quasi-CRM approach.

(3) Multiple toxicities

The aforementioned approaches collapse different types and/or grades of toxicity into a single numerical index (e.g., TTB or TTP) to summarize the overall severity of multiple toxicities. An alternative approach is to retain the original scale of the toxicity grade and use multiple toxicity constraints to quantify the safety of the drug (Lee et al., 2010). For example, the target dose may be defined as the dose with the grade 2 toxicity rate $\leq 40\%$, and the grade 3 and higher toxicity rate $\leq 30\%$. Lee et al. (2010) extended CRM to account for multiple toxicity constraints (referred to as the MC-CRM design hereafter) based on the proportional odds model. This MC-CRM design was further extended to trials with late-onset outcomes (Lee et al., 2017).

Among the above three approaches, we generally favor approaches 2 and 3. The difficulty of approach 1 is that it is challenging to define the target TTB (NETS, TBS, and TTP) in practice, as these composite scores lack intuitive clinical interpretation. For example, TTB = 3 does not have direct interpretation in terms of the toxicity rates in grades and types. In addition, different toxicity profiles may result in the same TTB, but have dramatically different clinical implications. Bekele and Thall (2004) and Ezzalfani et al. (2013) provided some guidance to facilitate the elicitation of the target, but specifying the target remains complicated and extremely challenging in practice. In contrast the elicitation of the target for approaches 2 and 3 is similar to that of conventional DLT-based dose finding designs.

The majority of designs accounting for toxicity grades and types are model-based. Albeit using different metrics to summarize toxicity, these designs all employ a similar strategy to CRM. That is, they assume a model to describe the relationship between toxicity grade/type and the dose; then, based on the accumulating data, they continuously update the model estimate and make the decision of dose assignment for the incoming new patient (typically by assigning the new patient to the dose for which the estimate of the toxicity metric is closest to the target) until the stopping rule is satisfied (e.g., the maximum sample size is reached). In order to account for toxicity grade and type, the model used by the designs is significantly more complex than that of CRM, e.g., the multinomial model for multiple-level toxicity grade (Lee et al., 2010), and the joint ordinal model to account for the correlation among toxicity types (Bekele and Thall, 2004). This makes these designs difficult to understand and implement, and thus they are rarely used in practice.

In this chapter, we describe two model-assisted designs that are transparent and easy to implement. The first one is the generalized BOIN (gBOIN) design, which accommodates continuous and quasi-binary toxicity endpoints, as well as the standard binary endpoint (Mu et al., 2018). The second method is the multiple-toxicity BOIN (MT-BOIN) design based on multiple toxicity constraints (Lin, 2018). Unlike model-based designs, the decision rules of the gBOIN and MT-BOIN designs can be tabulated prior to the onset of the trial, making the trial conduct simple and straightforward, similar to the standard BOIN design (with a binary endpoint).

7.2   gBOIN accounting for toxicity grade

7.2.1   Trial design

Consider a phase I trial with J prespecified doses, $d_1 < \cdots < d_J$. Let y denote the toxicity endpoint of interest, either a binary or quasi-binary outcome (e.g., DLT or ETS) or a continuous outcome (e.g., TTB, TBS, or TTP), belonging to the exponential family of distributions,

$$f(y|d_j) = h(y)\exp\{\eta(\theta_j)T(y) - A(\theta_j)\}, \tag{7.1}$$

where $h(\cdot)$, $T(\cdot)$, $\eta(\cdot)$, and $A(\cdot)$ are known functions, and θ_j is a distributional parameter that can be scalar or vector, depending on the distribution of y at dose d_j. The exponential family of distributions includes many commonly used distributions, such as normal, binomial, multinomial, Poisson, gamma, and beta distributions. For example, define $\mu = E(y)$ and $\mu_j = E(y|d_j)$, then

- y follows a Bernoulli distribution if $\theta_j = \mu_j$, $\eta(\theta_j) = \log\{\mu_j/(1-\mu_j)\}$, $A(\theta_j) = -\log(1-\mu_j)$, $T(y) = y$, and $h(y) = 1$.

- y follows a normal distribution if $\theta_j = \mu_j$, $\eta(\theta_j) = \mu_j/\sigma_j^2$, $A(\theta_j) = \mu_j^2/(2\sigma_j^2)$, $T(y) = y$, and $h(y) = (2\pi\sigma_j^2)^{-1/2}\exp\{-y^2/(2\sigma_j^2)\}$.

Let ϕ denote the target value of μ for dose finding. For binary or quasi-binary toxicity endpoints (DLT or ETS), ϕ is simply the target DLT probability; for continuous endpoints (e.g., the TTB, TBS, or TTP), ϕ is the target value of the TTB, TBS, or TTP. Let $D_j = (y_1, \cdots, y_{n_j})$ denote the observed toxicity data from n_j patients treated at dose d_j, and define the corresponding sample mean

$$\hat{\mu}_j = \sum_{i=1}^{n_j} y_i/n_j.$$

For a binary or quasi-binary toxicity endpoint (DLT or ETS), $\hat{\mu}_j$ is the observed toxicity rate at dose level j; and for continuous endpoints such as the TTB or TTP, $\hat{\mu}_j$ is the sample mean of the observed TTB or TTP at dose level j.

The gBOIN design shares a similarly concise dose escalation/de-escalation rule as the standard BOIN design (with a binary endpoint). Let λ_e and λ_d denote the optimal dose escalation/de-escalation boundaries. The gBOIN is summarized as follows:

1. Patients in the first cohort are treated at the lowest dose d_1, or the physician-specified dose.

2. Suppose j is the current dose level. To assign a dose to the next cohort of patients:

    - If $\hat{\mu}_j \leq \lambda_e$, escalate the dose to level $j + 1$.
    - If $\hat{\mu}_j > \lambda_d$, de-escalate the dose to level $j - 1$.
    - Otherwise, (i.e., $\lambda_e < \hat{\mu}_j \leq \lambda_d$), stay at the current dose level j.

3. Repeat Step 2 until the maximum sample size N is reached. At that point, select MTD as the dose whose isotonic estimate $\tilde{\mu}_j$ is closest to target ϕ, where $\{\tilde{\mu}_j, j = 1, \ldots, J\}$ is obtained by applying the pooled adjacent violators algorithm (Barlow et al., 1972) to $\{\hat{\mu}_j, j = 1, \ldots, J\}$.

Table 7.2 provides examples of the values of (λ_e, λ_d) for different target values of ϕ. The general formula to calculate (λ_e, λ_d) is described in the next section.

For patient safety, the following overdose control rule is imposed when using the gBOIN design:

If $\Pr(\mu_j > \phi \mid D_j) > 0.95$ and $n_j \geq 3$, dose level j and higher are eliminated from the trial, and the trial is terminated with no dose being selected as MTD if the lowest dose level is eliminated.

TABLE 7.2: Dose escalation and de-escalation boundaries (λ_e, λ_d) for Bernoulli, quasi-Bernoulli, and continuous toxicity endpoints.

	Bernoulli or quasi-Bernoulli endpoint (e.g., DLT or ETS)						
	Target toxicity probability ϕ						
	0.10	0.15	0.20	0.25	0.30	0.35	0.40
λ_e	0.078	0.118	0.157	0.197	0.236	0.276	0.316
λ_d	0.119	0.179	0.238	0.298	0.358	0.419	0.479

	Continuous endpoint (e.g., TTB, TBS or TTP)						
	Target toxicity value ϕ						
	0.25	0.5	0.75	1.0	1.5	2.0	3.0
λ_e	0.2	0.4	0.6	0.8	1.2	1.6	1.4
λ_d	0.3	0.6	0.9	1.2	1.8	2.4	3.6

*Under the default setting $\phi_1 = 0.6\phi$ and $\phi_2 = 1.4\phi$.

The posterior probability $\Pr(\mu_j > \phi \mid D_j) > 0.95$ can be evaluated on the basis of a beta-binomial model for the binary or quasi-binary endpoint, assuming μ_j follows a vague beta prior, e.g., $\mu_j \sim \text{Beta}(1,1)$, leading to a posterior $\mu_j | y_j, n_j \sim \text{Beta}(y_j + 1, n_j - y_j + 1)$. For normal endpoint y with mean μ_j and variance σ_j^2, assuming noninformative prior $(\mu, \sigma_j^2) \propto \sigma_j^{-2}$, the marginal posterior distribution of μ_j follows a student's t distribution with degrees of freedom $(n_j - 1)$, mean $\hat{\mu}_j$ and scale parameter $n_j^{-1} \sum_{i=1}^{n_j} (y_i - \hat{\mu}_j)^2$.

7.2.2 Statistical derivation and properties

The derivation of the gBOIN design is similar to that of the BOIN design, described in Section 3.4.2. To proceed, a class of nonparametric designs with dose escalation and de-escalation boundaries $(\lambda_e(d_j, n_j, \phi), \lambda_d(d_j, n_j, \phi))$ are introduced as follows:

1. Patients in the first cohort are treated at the lowest dose level or the physician-specified level.

2. Suppose j is the current dose level. To assign a dose to the next cohort of patients, consider the following:

   - If $\hat{\mu}_j \leq \lambda_e(d_j, n_j, \phi)$, escalate the dose level to $j + 1$.

   - If $\hat{\mu}_j > \lambda_d(d_j, n_j, \phi)$, de-escalate the dose level to $j - 1$.

   - Otherwise, i.e., $\lambda_e(d_j, n_j, \phi) < \hat{\mu}_j \leq \lambda_d(d_j, n_j, \phi)$, stay at the current dose level j.

3. Continue to repeat Step 2 until the maximum sample size N is reached.

As the dose escalation and de-escalation boundaries $\lambda_e(d_j, n_j, \phi)$ and $\lambda_d(d_j, n_j, \phi)$ are unspecified functions of dose d_j and n_j (i.e., the number of patients who have been treated at dose level j), this family of designs includes all possible nonparametric designs that do not impose a parametric model on the dose–toxicity curve and use the local data observed from the current dose level in decision making.

The gBOIN design is obtained by choosing the values of $\lambda_e(d_j, n_j, \phi)$ and $\lambda_d(d_j, n_j, \phi)$ that minimize the probability of making incorrect decisions on dose escalation and de-escalation. Toward that goal, following the approach of Liu and Yuan (2015), Mu et al. (2018) considered the following three point hypotheses:

$$H_{0j} : \mu_j = \phi, \quad H_{1j} : \mu_j = \phi_1, \quad H_{2j} : \mu_j = \phi_2,$$

where ϕ_1 indicates that the dose is substantially underdosing (i.e., below MTD) such that escalation is required, and ϕ_2 indicates that the dose is substantially overdosing such that de-escalation is required. Let $\mathcal{S}$, $\mathcal{E}$, and $\mathcal{D}$

denote stay (at the current dose), escalation, and de-escalation, respectively. Under H_{0j}, the correct decision is $\mathcal{S}$, and incorrect decisions are $\bar{\mathcal{S}} = \{\mathcal{E}, \mathcal{D}\}$; under H_{1j}, the correct decision is $\mathcal{E}$, and incorrect decisions are $\bar{\mathcal{E}} = \{\mathcal{S}, \mathcal{D}\}$; and under H_{2j}, the correct decision is $\mathcal{D}$, and incorrect decisions are $\bar{\mathcal{D}} = \{\mathcal{S}, \mathcal{E}\}$.

Taking a non-informative approach, the investigational dose d_j is assumed *a priori* to have an equal chance of being below, equal to, or above the target, i.e., $\Pr(H_{0j}) = \Pr(H_{1j}) = \Pr(H_{2j}) = 1/3$. Then the probability of making an incorrect decision, denoted by α, is given by

$$\alpha = \Pr(H_{0j})\Pr(\bar{\mathcal{S}}|H_{0j}) + \Pr(H_{1j})\Pr\left(\bar{\mathcal{E}}|H_{1j}\right) + \Pr(H_{2j})\Pr\left(\bar{\mathcal{D}}|H_{2j}\right)$$

$$= \frac{1}{3}\Pr\{\hat{\mu}_j \le \lambda_e\left(d_j, n_j, \phi\right) \text{ or } \hat{\mu}_j > \lambda_d\left(d_j, n_j, \phi\right)\} + \frac{1}{3}\Pr\{\hat{\mu}_j > \lambda_e\left(d_j, n_j, \phi\right)\}$$

$$+ \frac{1}{3}\Pr\{\hat{\mu}_j \le \lambda_d\left(d_j, n_j, \phi\right)\}. \tag{7.2}$$

Mu et al. (2018) derived the optimal values of $\lambda_e\left(d_j, n_j, \phi\right)$ and $\lambda_d\left(d_j, n_j, \phi\right)$ that minimize the decision error (7.2).

Theorem 7.1 *Let ϑ_k denote the model parameters under H_{kj}, $k = 0, 1, 2$. The probability of making an incorrect decision is minimized by*

$$\lambda_e = \frac{A(\vartheta_1) - A(\vartheta_0)}{\eta(\vartheta_1) - \eta(\vartheta_0)}, \quad \lambda_d = \frac{A(\vartheta_2) - A(\vartheta_0)}{\eta(\vartheta_2) - \eta(\vartheta_0)}, \tag{7.3}$$

given that the non-informative prior $\Pr(H_{0j}) = \Pr(H_{1j}) = \Pr(H_{2j}) = 1/3$ *is used.*

Specifically, when y is a binary or quasi-binary toxicity endpoint (DLT or ETS), we have $\vartheta_k = \phi_k$, $A(\vartheta_k) = -\log(1 - \phi_k)$, $\eta(\vartheta_k) = \log\{\phi_k/(1 - \phi_k)\}$, with $\phi_0 \equiv \phi$. Then,

$$\lambda_e = \frac{\log\left(\dfrac{1 - \phi_1}{1 - \phi}\right)}{\log\left(\dfrac{\phi(1 - \phi_1)}{\phi_1(1 - \phi)}\right)}, \quad \lambda_d = \frac{\log\left(\dfrac{1 - \phi}{1 - \phi_2}\right)}{\log\left(\dfrac{\phi_2(1 - \phi)}{\phi(1 - \phi_2)}\right)}, \tag{7.4}$$

which are the same as the boundaries of the BOIN design (Liu and Yuan, 2015) for a standard binary toxicity endpoint.

When y is a continuous endpoint (e.g., TTB, NETS, TBS, and TTP) following a normal distribution, we have $\vartheta_k = \phi_k$, $A(\vartheta_k) = \phi_k^2/(2\sigma_j^2)$, $\eta(\vartheta_k) = \phi_k/\sigma_j^2$. Then,

$$\lambda_e = \frac{\phi + \phi_1}{2}, \quad \lambda_d = \frac{\phi + \phi_2}{2}. \tag{7.5}$$

Therefore, gBOIN generalizes the BOIN design to embrace various types of endpoints.

Table 7.2 shows examples of the values of (λ_e, λ_d) for different target values of ϕ with $\phi_1 = 0.6\phi$ and $\phi_2 = 1.4\phi$. It is remarkable that, regardless the type of endpoint, the optimal dose escalation and de-escalation boundaries (λ_e, λ_d) are independent of d_j and n_j, which means that the same pair of boundaries can be used throughout the trial no matter which dose is the current dose or how many patients have been treated at the current dose. This feature makes the gBOIN design simple to implement in practice.

Similar to the BOIN design, the gBOIN design has the following desirable finite-sample and large-sample properties:

Theorem 7.2 *The gBOIN design is long-term memory coherent in the sense that the design will never escalate the dose when $\hat{\mu}_j > \phi$, and it will never de-escalate the dose when $\hat{\mu}_j < \phi$.*

Theorem 7.3 *As the number of patients goes to infinity, the dose assignment and the selection of MTD under the gBOIN design converge almost surely to dose level j^* if dose level j^* is the only dose satisfying $\mu_{j^*} \in (\lambda_e, \lambda_d)$. If there are multiple dose levels in (λ_e, λ_d), the design will converge almost surely to one of these levels.*

7.3   Multiple toxicity BOIN

The gBOIN design requires specifying a single toxicity endpoint (e.g., ETS, TTB, or TBS) to summarize the toxicity profile of the dose. In some cases, investigators prefer to retain the original scale of the toxicity grade (e.g., as scored by CTCAE) and use multiple toxicity constraints to identify MTD. For example, if the investigators are interested in finding MTD as the highest dose with Pr(grade 2 toxicity) $\leq 40\%$ and Pr(grade 3 and higher toxicity) $\leq 30\%$. The multiple-toxicity BOIN (MT-BOIN) design (Lin, 2018) can be used to achieve this goal.

7.3.1   Trial design

Consider a phase I trial with two binary toxicity endpoints of interest, Y_1 and Y_2, where, for example, $Y_1 = 1$ denotes the occurrence of grade 2 toxicity, and $Y_2 = 1$ denotes the occurrence of grade 3 or higher toxicity. Let $\pi_{lj} = \Pr(Y_l = 1|d_j)$ denote the toxicity probability of Y_l at dose d_j, with $\pi_{l1} < \cdots < \pi_{lJ}$, $l = 1, 2$, $d_j \in \{d_1, \ldots, d_J\}$. Lin (2018) discussed MT-BOIN designs for two different structures of Y_1 and Y_2, i.e., Y_1 and Y_2 are nested or non-nested. An example of Y_2 nested in Y_1 is that Y_1 denotes grade 2 or higher toxicity, and Y_2 denotes grade 3 or higher toxicity. Here, we focus on the case that Y_1 and Y_2 are not nested, e.g., Y_1 denotes grade 2 toxicity, and Y_2 denotes grade 3 or higher toxicity. In many cases, nested endpoints can be transformed into non-nested endpoints.

Let $\phi^{(l)}$ denote the target toxicity probability of Y_l, $l = 1, 2$. In the example above, $\phi^{(1)} = 40\%$ (for Y_1: grade 2 toxicity) and $\phi^{(2)} = 30\%$ (for Y_2: grade 3

TABLE 7.3: The escalation/de-escalation boundaries (λ_e, λ_d) under the BOIN design for different target toxicity rates ϕ, using the default underdosing toxicity probability $\phi_1 = 0.6\phi$ and overdosing toxicity probability $\phi_2 = 1.4\phi$.

Boundaries	Target toxicity probability ϕ					
	0.15	0.2	0.25	0.3	0.35	0.4
λ_e	0.118	0.157	0.197	0.236	0.276	0.316
λ_d	0.179	0.238	0.298	0.358	0.419	0.479

or higher toxicity). With the two toxicity constraints, MTD (i.e., target dose) is defined as

$$d_{j\dagger} = \min\{d_{j_1^*}, d_{j_2^*}\}, \tag{7.6}$$

where

$$d_{j_l^*} = \underset{d_j \in \{d_1,\dots,d_J\}}{\arg\min} \; |\pi_{lj} - \phi^{(l)}|, \quad l = 1, 2.$$

Here, $d_{j_l^*}$ is the dose that has the toxicity probability closest to the target toxicity probability with respect to Y_l. Depending on trial objectives, other definitions of MTD can also be used when appropriate.

Suppose that at dose d_j, a total of y_{lj} out of n_j patients have experienced the toxicity event associated with Y_l. The observed toxicity rate of Y_l at d_j is

$$\hat{\pi}_{lj} = y_{lj}/n_j, \quad l = 1, 2, \; j = 1, \dots, J.$$

Let $(\lambda_e^{(1)}, \lambda_d^{(1)})$ denote the optimal dose escalation/de-escalation boundaries corresponding to Y_1, which are the same as those of the standard BOIN design with Y_1 as the endpoint and $\phi^{(1)}$ as the target. Similarly, let $(\lambda_e^{(2)}, \lambda_d^{(2)})$ denote the dose escalation/de-escalation boundaries corresponding to Y_2, which are the same as those of the standard BOIN design with Y_2 as the endpoint and $\phi^{(2)}$ as the target. Table 7.3 provides the standard BOIN dose escalation/de-escalation boundaries for different targets, see Section 3.4 for more details. In the above trial example, $(\lambda_e^{(1)}, \lambda_d^{(1)}) = (0.316, 0.479)$ and $(\lambda_e^{(2)}, \lambda_d^{(2)}) = (0.236, 0.358)$.

The MT-BOIN design has a similarly concise dose escalation/de-escalation rule as the standard BOIN design (with a binary endpoint), by simply comparing the observed toxicity rate $\hat{\pi}_{lj}$ with $\lambda_e^{(l)}$ and $\lambda_d^{(l)}$, $l = 1, 2$. The MT-BOIN design is illustrated in Figure 7.1 and described as follows:

1. Patients in the first cohort are treated at the lowest dose d_1, or the physician-specified dose.

2. Suppose j is the current dose level. To assign a dose to the next cohort of patients:

   - If $\hat{\pi}_{1j} \leq \lambda_e^{(1)}$ and $\hat{\pi}_{2j} \leq \lambda_e^{(2)}$, escalate the dose to level $j + 1$.

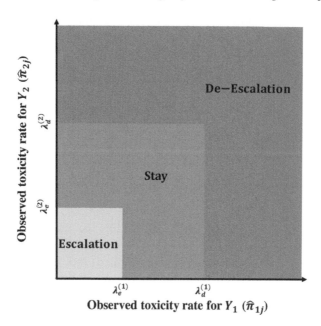

FIGURE 7.1: Dose escalation/retainment/de-escalation regions of the MT-BOIN design with two toxicity endpoints Y_1 and Y_2, where $\lambda_e^{(l)}$ and $\lambda_d^{(l)}$ are standard BOIN dose escalation and de-escalation boundaries corresponding to Y_l, $l = 1, 2$.

- If $\hat{\pi}_{1j} > \lambda_d^{(1)}$ or $\hat{\pi}_{2j} > \lambda_d^{(2)}$, de-escalate the dose to level $j - 1$.
- Otherwise, i.e., $\hat{\pi}_{1j} \leq \lambda_d^{(1)}$, $\lambda_e^{(2)} < \hat{\pi}_{2j} \leq \lambda_d^{(2)}$ or $\lambda_e^{(1)} < \hat{\pi}_{1j} \leq \lambda_d^{(1)}$, $\hat{\pi}_{2j} \leq \lambda_d^{(2)}$, stay at the current dose level j.

3. Repeat Step 2 until the maximum sample size N is reached. At that point, perform the isotonic regression (Barlow et al., 1972) to the estimated toxicity rates $\{\hat{\pi}_{lj}, j = 1, \ldots, J\}$, so that the isotonically transformed estimator $\{\tilde{\pi}_{lj}, j = 1, \ldots, J\}$ satisfies the monotonic constraint, and then select the overall MTD $\hat{j}^\dagger$ as

$$d_{\hat{j}^\dagger} = \min\{d_{\hat{j}_1^*}, d_{\hat{j}_2^*}\},$$

where $d_{\hat{j}_l^*} = \arg\min\limits_{d_j \in \mathcal{N}} |\tilde{\pi}_{lj} - \phi^{(l)}|$, and the set $\mathcal{N} = \{d_j : n_j > 0\}$ contains all the doses that have been tested in the trial.

For the purpose of overdose control, the MT-BOIN design imposes a dose elimination rule similar to BOIN as follows:

If any of $\Pr(\pi_{lj} > \phi^{(l)} \mid D_j) > 0.95$, $l = 1, 2$, and $n_j \geq 3$, dose level j and higher are eliminated from the trial, and the trial is terminated if the lowest dose level is eliminated.

The value of $\Pr(\pi_{lj} > \phi \mid D_j)$ is evaluated based on the beta-binomial model (see Section 3.4). When the trial is terminated due to toxicity, no dose should be selected as MTD.

7.3.2 Statistical derivation and properties

Let $\phi_1^{(l)} < \phi^{(l)}$ denote the highest probability considered to be subtherapeutic such that dose escalation is needed, and let $\phi_2^{(l)} > \phi^{(l)}$ be the lowest value deemed excessively toxic such that dose de-escalation is warranted, $l = 1, 2$. Following Liu and Yuan (2015), three hypotheses are considered for the MT-BOIN design with non-nested outcomes,

$$H_{0j} : (\pi_{1j}, \pi_{2j}) \in \Theta_{0j}, \quad H_{1j} : (\pi_{1j}, \pi_{2j}) \in \Theta_{1j}, \quad H_{2j} : (\pi_{1j}, \pi_{2j}) \in \Theta_{2j}, \tag{7.7}$$

where the parameter set Θ_{ij}, $i = 0, 1, 2$ is defined as

$$\Theta_{0j} = \left\{ (\phi^{(1)}, \phi^{(2)}), (\phi^{(1)}, \phi_1^{(2)}), (\phi_1^{(1)}, \phi^{(2)}) \right\},$$

$$\Theta_{1j} = \left\{ (\phi_1^{(1)}, \phi_1^{(2)}) \right\},$$

$$\Theta_{2j} = \left\{ (\phi_2^{(1)}, \phi_2^{(2)}), (\phi_2^{(1)}, \phi^{(2)}), (\phi_2^{(1)}, \phi_1^{(2)}), (\phi^{(1)}, \phi_2^{(2)}), (\phi_1^{(1)}, \phi_2^{(2)}) \right\}.$$

Unlike the standard BOIN design that only involves three point hypotheses, each hypothesis H_{ij}, $i = 0, 1, 2$ of the proposed MT-BOIN enjoys a hierarchical structure and includes a set of multiple point sub-hypotheses, which is thus composite. In particular, Θ_{1j} indicates that both toxicity rates for Y_1 and Y_2 at dose level j are overly small, hence, the hypothesis H_{1j} corresponds to the subtherapeutic case. Θ_{2j} is made up by the combinations that at least one toxicity rate is excessively high, then H_{2j} means that dose level j is not safe. Similarly, H_{0j} indicates that the dose is within the target zone. Note that under H_{0j}, the dose level with the joint toxicity probabilities $(\pi_{1j}, \pi_{2j}) = (\phi^{(1)}, \phi_1^{(2)})$ (or $(\pi_{1j}, \pi_{2j}) = (\phi_1^{(1)}, \phi^{(2)})$) is treated as proper dosing instead of underdosing. This is because the monotone increasing dose–toxicity relationship implies that $\phi^{(1)} < \pi_{1,j+1}$ (or $\phi^{(2)} < \pi_{2,j+1}$); as a result, dose level $j + 1$ might be overly toxic in terms of toxicity Y_1 (or Y_2).

Let $O_j = (y_{1j}, y_{2j}, n_j)$ be the local data observed at dose level j. Denote the prior model probability $\Pr(H_{ij})$ by ω_{ij}. For simplicity, the uniform prior $\omega_{ij} = 1/3$ is considered, $i = 0, 1, 2$. The correct dose-assignment decisions under H_{0j}, H_{1j}, and H_{2j} are $\mathcal{S}$, $\mathcal{E}$, and $\mathcal{D}$, respectively. To account for the composite nature of the three hypotheses, Lin (2018) defined $\alpha_{\max}$, the maximum probability of making incorrect decisions with multiple toxicity outcomes, as follows:

$$
\begin{aligned}
\alpha_{\max} &= \omega_{1j}\Pr(\mathcal{S}\text{ or }\mathcal{D}\mid H_{1j}) + \omega_{0j}\Pr(\mathcal{D}\text{ or }\mathcal{E}\mid H_{0j}) + \omega_{2j}\Pr(\mathcal{E}\text{ or }\mathcal{S}\mid H_{2j}) \\
&= \sum_{y_{1j}=0}^{n_j}\sum_{y_{2j}=0}^{n_j}\Bigg[\omega_{1j}I(\hat\pi_{1j} > \lambda_e^{(1)}\text{ or }\hat\pi_{2j} > \lambda_e^{(2)})\max_{(\pi_{1j},\pi_{2j})\in\Theta_{1j}} f(O_j;\pi_{1j},\pi_{2j}) \\
&\quad + \omega_{0j}I(\hat\pi_{1j}\le\lambda_e^{(1)}\text{ and }\hat\pi_{2j}\le\lambda_e^{(2)})\max_{(\pi_{1j},\pi_{2j})\in\Theta_{0j}} f(O_j;\pi_{1j},\pi_{2j}) \\
&\quad + \omega_{0j}I(\hat\pi_{1j} > \lambda_d^{(1)}\text{ or }\hat\pi_{2j} > \lambda_d^{(2)})\max_{(\pi_{1j},\pi_{2j})\in\Theta_{0j}} f(O_j;\pi_{1j},\pi_{2j}) \\
&\quad + \omega_{2j}I(\hat\pi_{1j}\le\lambda_d^{(1)}\text{ and }\hat\pi_{2j}\le\lambda_d^{(2)})\max_{(\pi_{1j},\pi_{2j})\in\Theta_{2j}} f(O_j;\pi_{1j},\pi_{2j})\Bigg],
\end{aligned}
$$

where $f(O_j;\pi_{1j},\pi_{2j})$ is the joint binomial likelihood function for Y_1 and Y_2. Due to the small sample size of the phase I dose-finding trials, the incorporation of the correlation between the two toxicity outcomes does not necessarily improve the performance of the design. Therefore, the joint likelihood $f(O_j;\pi_{1j},\pi_{2j})$ can be simply treated as the product of independent binomial likelihood functions. Note that even when the two outcomes are correlated, the marginal estimate of the toxicity rate under the working independence assumption is still consistent.

Based on the minimax theory, Lin (2018) obtained explicit expressions of optimal interval boundaries $\lambda_e^{(l)}(d_j, n_j, \phi^{(l)})$ and $\lambda_d^{(l)}(d_j, n_j, \phi^{(l)})$ for non-nested toxicities by minimizing $\alpha_{\max}$.

Theorem 7.4 *If uniform prior model probabilities are considered, the optimal interval boundaries obtained by minimizing the maximum incorrect probability (7.8) for the toxicity Y_l are exactly the same as those of standard BOIN by treating Y_l alone, that is, for $l = 1, 2$,*

$$
\lambda_e^{(l)} = \frac{\log\left(\dfrac{1-\phi_1^{(l)}}{1-\phi^{(l)}}\right)}{\log\left(\dfrac{\phi^{(l)}(1-\phi_1^{(l)})}{\phi_1^{(l)}(1-\phi^{(l)})}\right)}, \qquad
\lambda_d^{(l)} = \frac{\log\left(\dfrac{1-\phi^{(l)}}{1-\phi_2^{(l)}}\right)}{\log\left(\dfrac{\phi_2^{(l)}(1-\phi^{(l)})}{\phi^{(l)}(1-\phi_2^{(l)})}\right)}.
\tag{7.8}
$$

The proof of Theorem 7.4 is provided in Lin (2018). Remarkably, Theorem 7.4 shows that the optimal boundaries of MT-BOIN coincide exactly with those of standard BOIN based on a single toxicity outcome and are independent of d_j and n_j. Because the optimal interval boundaries from (7.8) minimize the maximum probability of incorrect decisions, MT-BOIN enjoys optimality within the aforementioned class of local-data designs for multiple toxicities.

7.4 Software and illustration

Software for the gBOIN and MT-BOIN designs are available as a module of "BOIN Suite" at http://www.trialdesign.org. In what follows, we use the solid tumor trial introduced in Section 3.6 to illustrate the use of gBOIN to design the trial. Recall that the objective of the trial is to determine MTD for the MDM2/MDMX inhibitor ALRN-6924 in combination with paclitaxel in adult patients with advanced or metastatic solid tumors. Five doses of ALRN-6924 are investigated with the target DLT probability $\phi = 25\%$. For the purpose of illustration, we here consider an alternative target DLT probability $\phi = 30\%$. For MT-BOIN, design parameters and setup are similar to these of the standard BOIN design, see Section 3.6 for details.

Suppose that investigators anticipate that the agent may induce a high percentage of low grade toxicities that may cause substantial dose reduction and discontinuation. Thus, the investigators prefer to account for low grade toxicities in the decisions of dose escalation/de-escalation and MTD determination. After selecting and launching the "gBOIN/MT-BOIN" module from the BOIN Suite launchpad, the trial can be designed using the following three steps:

Step 1: Enter trial parameters
Doses and Sample Size   As shown in Figure 7.2, the number of doses under investigation is five; the starting dose level is one. The cohort size for the trial is three and the number of cohorts is 10, with a total sample size of 30. As a rule of thumb, we recommend the maximum sample size $N = 6 \times J$ (i.e., the

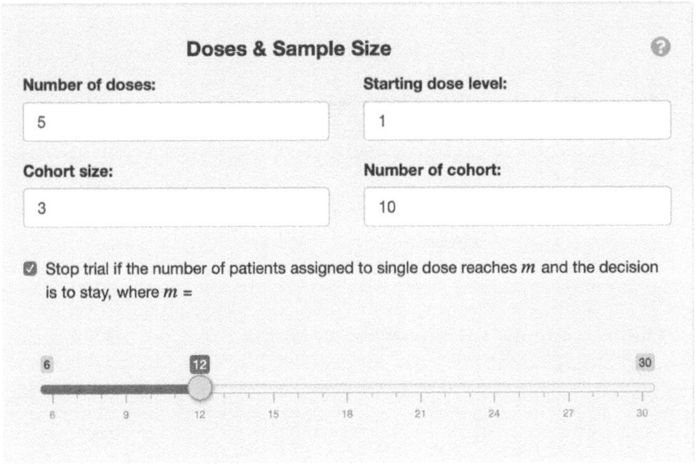

FIGURE 7.2: Specify doses, sample size, and convergence stopping rule.

maximum sample size of the 3+3 design) as the total sample size, where J is the number of doses. This sample size generally yields reasonable operating characteristics (e.g., 50-70% correct selection percentage of the true MTD). Section 3.6 provides more guidance on the determination of cohort size.

To reduce the sample size, it is often useful to use the "convergence" stopping rule: stop the trial early when m patients have been assigned to a dose and the decision is to stay at that dose. This stopping criterion suggests that the dose finding approximately converges to MTD, thus the trial can be stopped. We recommend $m = 9$ or larger. In this trial, $m = 12$ is used. Because of the early stopping rule, the actual sample size used in the trial is often smaller than the prespecified maximum sample size N. The saving depends on the true dose–toxicity scenario and can be evaluated using simulation. Usually, the saving in the sample size is more prominent when the true MTD is near the starting dose.

Target and Equivalent DLT   As shown in Figure 7.3, the target DLT probability is $\phi = 0.3$. When appropriate, the accelerated titration can be chosen to speed up dose escalation and reduce the total sample size. See Section 3.6 for more discussion on choosing the target and pros and cons of the accelerated titration.

For gBOIN, the new design parameters are "Equivalent Number of DLT," which maps toxicity grades to numeric scores that reflect their relative severity in the unit of DLT. This is the "Toxicity score" approach discussed in Section 7.1. For example, in Figure 7.3, grade 3 or higher toxicity is defined as a DLT; grade 2 toxicity is regarded as equivalent to 0.5 DLT, while grade 1 toxicity is regarded as acceptable and should not affect decisions of dose transition and MTD determination. The toxicity score should be elicited from clinicians and reflects the relative severity of different toxicity grades.

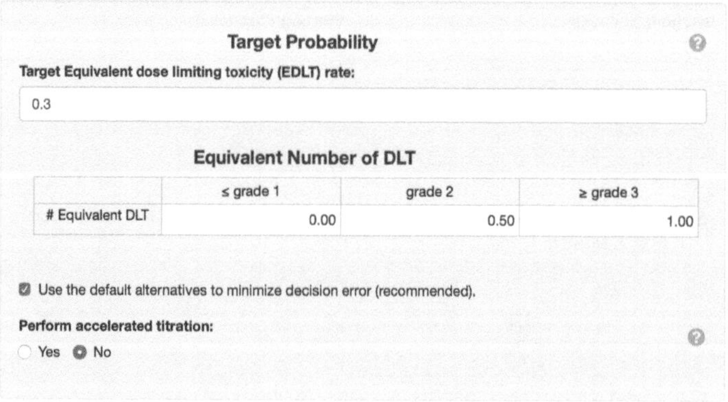

Target Probability

Target Equivalent dose limiting toxicity (EDLT) rate:

0.3

Equivalent Number of DLT

	≤ grade 1	grade 2	≥ grade 3
# Equivalent DLT	0.00	0.50	1.00

☑ Use the default alternatives to minimize decision error (recommended).

Perform accelerated titration:

○ Yes ● No

FIGURE 7.3: Specify the target and equivalent DLT scores.

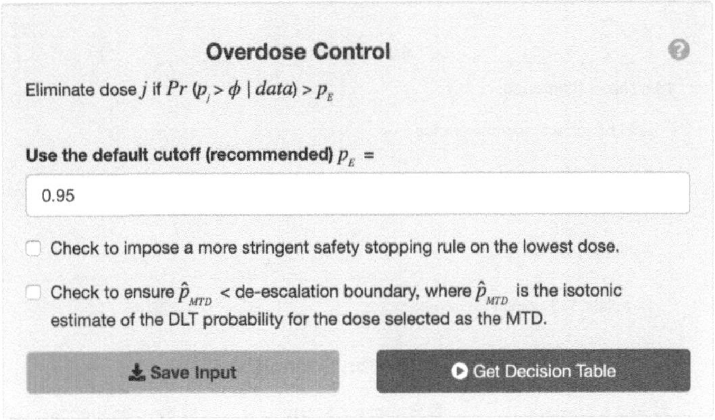

FIGURE 7.4: Specify the overdose control rules.

Overdose Control   This panel (see Figure 7.4) specifies the overdose control rule, and it is the same as that of standard BOIN. Section 3.6 provides guidance on how to set up the parameters. After completing the specification of trial parameters, the decision table will be generated by clicking the "Get Decision Table" button. The decision table will be automatically included in the protocol template in Step 3, but can also be saved as a separate csv, Excel and pdf file in this step if needed.

Step 2: Run simulation

Operating Characteristics   This step generates the operating characteristics of the design through simulation. As gBOIN considers multiple toxicity grades, the construction of scenarios is more complicated. Users need to provide the probability of each toxicity grade for each dose, see Figure 7.5. The scenarios used for simulation should cover various possible clinical scenarios, e.g., MTD is located at different dose levels. To facilitate the generation of the operating characteristics of the design, the software automatically provides a set of randomly generated scenarios with various MTD locations, which are often adequate for most trials. Depending on the application, users can modify, add or remove scenarios.

Figure 7.6 shows the operating characteristics of gBOIN for the solid tumor trial. The simulation results will be automatically included as a table in the protocol template in the next step, but can also be saved as a separate csv or Excel file if needed.

Step 3: Generate protocol template

Protocol Preparation   The gBOIN software generates sample texts and a protocol template (including the simulation results in Step 2) to facilitate

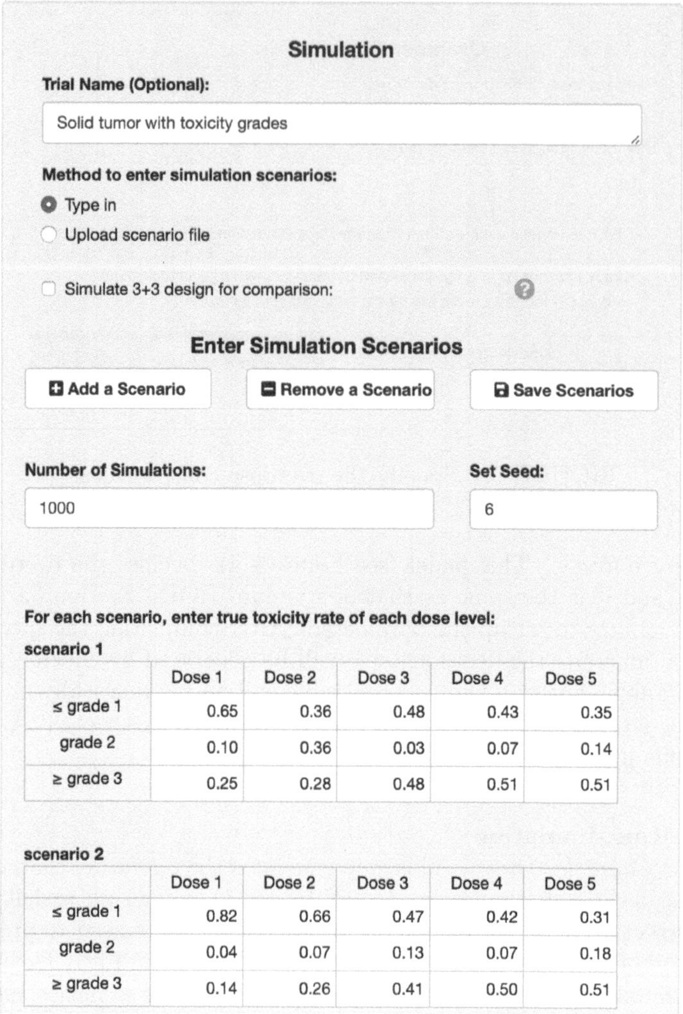

Simulation

Trial Name (Optional):

Solid tumor with toxicity grades

Method to enter simulation scenarios:
- ● Type in
- ○ Upload scenario file

☐ Simulate 3+3 design for comparison: ?

Enter Simulation Scenarios

| ➕ Add a Scenario | ➖ Remove a Scenario | 💾 Save Scenarios |

Number of Simulations: Set Seed:

| 1000 | 6 |

For each scenario, enter true toxicity rate of each dose level:

scenario 1

	Dose 1	Dose 2	Dose 3	Dose 4	Dose 5
≤ grade 1	0.65	0.36	0.48	0.43	0.35
grade 2	0.10	0.36	0.03	0.07	0.14
≥ grade 3	0.25	0.28	0.48	0.51	0.51

scenario 2

	Dose 1	Dose 2	Dose 3	Dose 4	Dose 5
≤ grade 1	0.82	0.66	0.47	0.42	0.31
grade 2	0.04	0.07	0.13	0.07	0.18
≥ grade 3	0.14	0.26	0.41	0.50	0.51

FIGURE 7.5: Simulate operating characteristics of the design.

the protocol write-up. The protocol template can be downloaded in various formats (e.g., html or Word file). Use of this module requires the completion of Steps 1 and 2. Once the protocol is approved by regulatory bodies (e.g., Institutional Review Board), we follow the design decision table to conduct the trial and make adaptive decisions (e.g., dose escalation/stay/de-escalation).

When the trial is completed, use the "Select MTD" module to select MTD. The software outputs the recommended MTD, the posterior estimate of DLT

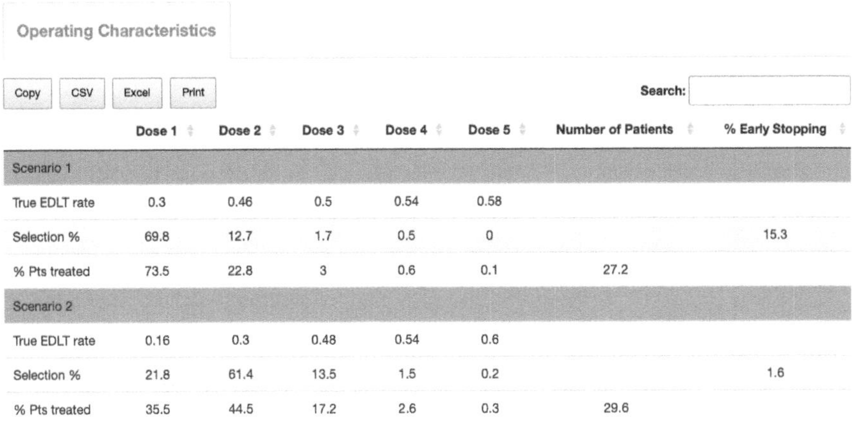

Operating Characteristics

| Copy | CSV | Excel | Print | | | | Search: | |

	Dose 1	Dose 2	Dose 3	Dose 4	Dose 5	Number of Patients	% Early Stopping
Scenario 1							
True EDLT rate	0.3	0.46	0.5	0.54	0.58		
Selection %	69.8	12.7	1.7	0.5	0		15.3
% Pts treated	73.5	22.8	3	0.6	0.1	27.2	
Scenario 2							
True EDLT rate	0.16	0.3	0.48	0.54	0.6		
Selection %	21.8	61.4	13.5	1.5	0.2		1.6
% Pts treated	35.5	44.5	17.2	2.6	0.3	29.6	

FIGURE 7.6: Operating characteristics of the gBOIN design for the solid tumor trial.

probability at each dose, and a 95% credible interval (see Figure 7.7). Given the data entered, dose level three is selected as MTD. In contrast, if we considered only DLT, ignoring grade 2 toxicities, then dose level four or five would be selected as MTD.

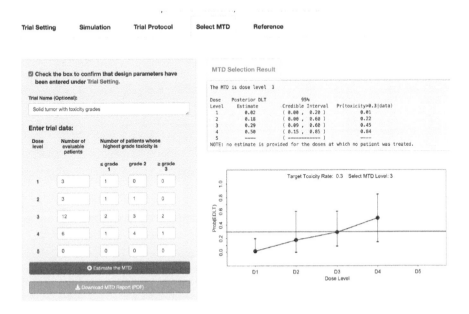

FIGURE 7.7: MTD estimation at the completion of the solid tumor trial based on a hypothetical dataset.

8

Finding Optimal Biological Dose

8.1 Introduction

Conventionally, the primary objective of phase I oncology trials is to establish the maximum tolerated dose (MTD). This more-is-better paradigm is based on the monotonicity assumption that a higher dose leads to higher toxicity and also higher efficacy, which typically holds for conventional chemotherapies.

The advent of novel targeted therapy and immunotherapy, such as checkpoint inhibitors and chimeric antigen receptor (CAR) T-cell therapy, has revolutionized cancer treatment. For these novel therapies, although toxicity typically increases with the dose, efficacy may plateau or even decrease at high doses (Cook et al., 2015; Sachs et al., 2016; Shah et al., 2021). In addition, some targeted therapies demonstrate minimal toxicity in the therapeutic dose range, making MTD unlikely to be reached (Mathijssen et al., 2014). In these cases, the conventional more-is-better paradigm, ignoring efficacy data, does not depict the underlying setting and may result in undesirable consequences. (Shah et al., 2021). As a result, FDA Oncology Center of Excellence recently initiated Project Optimus "to reform the dose optimization and dose selection paradigm in oncology drug development" (FDA, 2022).

To illustrate the issue, consider five doses of a targeted or immunotherapy agent, $d_1 < \cdots < d_5$. Suppose that the true toxicity probabilities for the five doses are $(\pi_{T1}, \ldots, \pi_{T5}) = (0.05, 0.12, 0.27, 0.35, 0.50)$. If the true efficacy probabilities are $(\pi_{E1}, \ldots, \pi_{E5}) = (0.20, 0.35, 0.36, 0.37, 0.38)$, then efficacy reaches a plateau of about 0.35 at dose d_2. All dose-finding methods considering toxicity only and with a target toxicity probability of 0.3 or 0.25, are most likely to select d_3 as MTD and use it for cohort expansion or a subsequent phase II trial. However, d_2 is obviously more desirable than MTD (i.e., d_3) with much lower toxicity and virtually identical efficacy. Any "toxicity-only" phase I method cannot determine this because it ignores efficacy.

Furthermore, suppose that the toxicity rates are the same but the true efficacy probabilities are $(\pi_{E1}, \ldots, \pi_{E5}) = (0.01, 0.05, 0.30, 0.60, 0.60)$. Escalating from d_3 to d_4 only increases the toxicity probability from 0.27 to 0.35, but doubles the efficacy probability from 0.3 to 0.6. Often this small increase in toxicity may be considered as a reasonable trade-off for the large increase in efficacy by choosing d_4 rather than d_3, however, toxicity-only methods cannot determine this.

DOI: 10.1201/9780429052781-8

Lastly, if the agent is ineffective for all doses, with true efficacy probabilities $(\pi_{E1}, \ldots, \pi_{E5}) = (0.00, 0.01, 0.01, 0.02, 0.02)$, the best decision is to not choose any dose, but the toxicity-based methods still are most likely to choose d_3. These examples demonstrate the deficiency of choosing a "best" dose based only on toxicity (e.g., MTD), ignoring efficacy, which may lead to the failure of subsequent phase II or III trials.

8.1.1   Phase I–II design paradigm

The aforementioned issues can be avoided by adopting the phase I–II design paradigm, which considers both toxicity and efficacy to adaptively allocate patients and identify the optimal biological dose (OBD). In general, the OBD is defined as the dose that optimizes a certain metric of the risk-benefit trade-off. Figure 8.1 shows the schema of phase I–II designs, which consist of the following elements:

- **Toxicity and efficacy outcomes** that characterize potential risks and benefits of the treatment being investigated. Examples of toxicity outcome include dose limiting toxicity (DLT) or tolerability rates, and examples of efficacy outcome include tumor responses, pharmacodynamic (PD) biomarkers measuring biological activities of the drug (e.g., inhibition of an oncogene pathway or proliferation of T cells), or surrogate efficacy endpoints (e.g., ctDNA, minimal residual disease). In this chapter, when describing designs, we use the toxicity rate and efficacy rate to generically represent the risk and benefit of the treatment, respectively, which should be customized based on the trial under consideration.

- **Risk–benefit trade-off criterion** that characterizes and quantifies the trade-off between efficacy and toxicity for each dose of the drug. Specific example includes the trade-off based on marginal toxicity and efficacy probabilities (Section 8.2.1) or utility based on the values of toxicity and efficacy endpoints (Section 8.3.1).

- **Optimal biological dose (OBD)** that maximizes the predefined risk–benefit trade-off criterion among all considered doses.

- **Statistical model** that describes the relationship between dose, toxicity, and efficacy. Examples include the Gumbel-Morgenstern copula model (Section 8.2.2) and multinomial model for bivariate binary toxicity and efficacy endpoints (Section 8.3.2).

- **Adaptive decision rule** that determines the best dose for the next cohort, based on the (dose, toxicity, efficacy) data observed from previous patients. Typically, the next cohort is assigned to the dose that has the highest estimate of the risk-benefit trade-off, based on the interim data.

- **Admissibility rules** that protect patients in the trial from unacceptably toxic or inefficacious doses. Admissibility rules are used to define admissible doses that satisfy certain prespecified toxicity and efficacy requirements.

Upon identification, only admissible doses can be used to treat patients. Examples will be provided in the next section.

- **Stopping rule** that terminates the trial early if all the doses being considered are unacceptably toxic or inefficacious.

- **Selection rule** that identifies OBD for subsequent studies if the trial is not terminated early. Typically, OBD is selected as the dose that is admissible and has the most favorable estimate of the risk-benefit trade-off.

A number of phase I–II designs have been proposed in recent years. The majority of them are model-based. These model-based designs consider different settings (e.g., different types of endpoints, statistical models, risk-benefit trade-off criteria, and decision rules), but share a similar design strategy to that outlined in Figure 8.1. Thall and Cook (2004) introduced a toxicity–efficacy trade-off contour for dose finding based on the Gumbel-Morgenstern copula model (hereafter referred to as the EffTox design). Bekele and Shen (2005) considered jointly modeling a binary toxicity outcome and a continuous efficacy outcome in phase I–II dose-finding oncology trials. Hunsberger et al. (2005) proposed a slope-sign design to guide dose escalation. Yin et al. (2006) parameterized both the toxicity and efficacy probabilities based on multivariate normal random variables, and they applied the global cross-ratio model (Dale, 1986) to jointly quantify the toxicity and efficacy outcomes. Zhang et al. (2006) developed a flexible continuation-ratio model to select the optimal dose at each decision-making time. Zang et al. (2014) used the logistic model to depict the local dose–efficacy relationship around the current dose. Liu and Johnson (2016) proposed a nonparametric design using a flexible Bayesian dynamic model for monotone dose–response curves. Riviere et al. (2018) utilized adaptive randomization to determine the plateau in phase I–II dose-finding designs. More sophisticated phase I–II designs were developed to handle more complex early-phase trials; for example, drug-combination trials (Yuan and

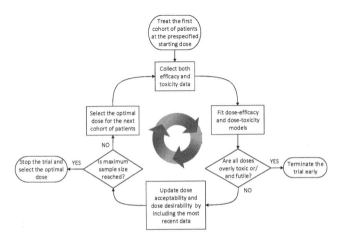

FIGURE 8.1: Diagram of phase I–II trial design.

Yin, 2011a; Cai et al., 2014; Guo and Li, 2015), dose–schedule optimization (Guo et al., 2016; Lin et al., 2020a, 2021), ordinal toxicity and efficacy outcomes (Houede et al., 2010; Lee et al., 2016), personalized dose-finding based on biomarkers (Guo and Yuan, 2017), and immunotherapy considering toxicity, efficacy and immune response (Liu et al., 2018), among others. For a comprehensive review on model-based phase I–II designs, see Mandrekar et al. (2010) and Yuan et al. (2016b).

These model-based phase I–II designs are statistically complicated and computationally intensive. To account for toxicity and efficacy, the model used by the designs is significantly more complex than that of the conventional phase I trial design (e.g., CRM), as illustrated in the next section. Highly complicated and structured parametric models also make the design susceptible to model misspecification. As a result, the model-based phase I–II designs are often regarded by practitioners as difficult to understand and implement, severely limiting their use in practice.

Model-assisted designs have been proposed to simplify the implementation of phase I–II trials, while yielding the performance comparable to or even better than model-based designs. As model-assisted designs do not assume any parametric dose-toxicity and -efficacy curves, they are also more robust than model-based designs. Lin and Yin (2017b) developed a toxicity-efficacy interval based phase I–II design (called STEIN) using the pool-adjacent-violators algorithm and model averaging. Li et al. (2017) proposed a toxicity and efficacy probability interval design that separately models toxicity and efficacy. Takeda et al. (2018) presented a Bayesian optimal interval design that accommodates both efficacy and toxicity endpoints, known as BOIN-ET. Zhou et al. (2019) developed a two-stage utility-based Bayesian optimal interval (U-BOIN) design using a Dirichlet-multinomial model to jointly model toxicity and efficacy. Lin et al. (2020b) adopted a quasi-binomial likelihood approach to model the observed utility directly, and developed the BOIN12 design. Shi et al. (2021) developed a utility-based toxicity probability interval (uTPI) design as an extension of the toxicity-only keyboard design (Yan et al., 2017).

In the following sections, we first introduce the EffTox design as an example to illustrate the characteristics of model-based designs, and then we describe several model-assisted phase I–II designs, including the BOIN12, U-BOIN, and uTPI designs. Other model-assisted designs such as STEIN and BOIN-ET will be briefly discussed at the end of this chapter.

8.2   EffTox design

The EffTox design is a model-based phase I–II design. It considers a binary toxicity endpoint y_T and a binary efficacy endpoint y_E, with $y_T = 1$ and $y_E = 1$ denoting the occurrence of toxicity and efficacy events, respectively. Let $d_1 < \cdots < d_J$ denote J prespecified doses. We assume that the (marginal) toxicity probability increases with the dose, i.e., $\pi_{T1} < \cdots < \pi_{TJ}$. The (marginal) efficacy probability $\pi_{E1}, \cdots, \pi_{EJ}$, however, does not necessarily increase with the dose, which may plateau or even decrease at higher doses. The EffTox design uses an efficacy–toxicity trade-off contour as the criterion to define and select OBD (Thall and Cook, 2004).

8.2.1 Efficacy–toxicity trade-off

To construct trade-off contours, we first elicit from clinicians three equally desirable efficacy–toxicity probability pairs $(\pi_E^{(1)}, 0)$, $(1, \pi_T^{(2)})$, and $(\pi_E^{(3)}, \pi_T^{(3)})$, subject to the constraints $\pi_E^{(1)} < \pi_E^{(3)}$ and $\pi_T^{(3)} < \pi_T^{(2)}$. For example, $(\pi_E^{(1)}, 0) =$ $(0.25, 0)$, $(1, \pi_T^{(2)}) = (1, 0.60)$, and $(\pi_E^{(3)}, \pi_T^{(3)}) = (0.40, 0.15)$ means that a dose with an efficacy rate of 25% and no toxicity, a dose with 100% efficacy and a toxicity rate of 60%, and a dose with an efficacy rate of 40% and a toxicity rate of 15% are considered equally desirable.

Based on the three equally desirable efficacy–toxicity probability pairs, an efficacy–toxicity trade-off (or utility) function is defined as a function of the marginal toxicity and efficacy probabilities

$$\psi(\pi_E, \pi_T) = 1 - \left\{ \left(\frac{\pi_E - 1}{\pi_E^{(1)} - 1} \right)^r + \left(\frac{\pi_T - 0}{\pi_T^{(2)} - 0} \right)^r \right\}^{1/r}, \qquad (8.1)$$

where $r > 0$ controls the degree of the curvature of the trade-off contours (Thall and Cook, 2004). The value of r is determined by solving the equation

$$\psi(\pi_E^{(3)}, \pi_T^{(3)}) = \psi(\pi_E^{(1)}, 0) = \psi(1, \pi_T^{(2)}).$$

Once the value of r is determined, the family of efficacy–toxicity trade-off contours is obtained by the utility function (8.1). Specifically, for a given utility value δ, the efficacy–toxicity trade-off contour is defined as

$$\mathcal{C}_\delta = \{(\pi_E, \pi_T) : \psi(\pi_E, \pi_T) = \delta\}.$$

That is, all (π_E, π_T) pairs in $\mathcal{C}_\delta$ have the same utility value δ.

Figure 8.2 gives an example for the efficacy–toxicity trade-off contours based on the function (8.1). The utility is standardized between 0 and 1, with the right bottom corner representing the most desirable case with a utility of 1, where efficacy is certain and toxicity is impossible, and the left upper corner representing the least desirable case with a utility of 0, where efficacy is impossible and toxicity is certain. All (π_E, π_T) pairs on each contour are equally desirable. The utilities of the contours increase moving from upper left to lower right, as π_T becomes smaller and π_E becomes larger.

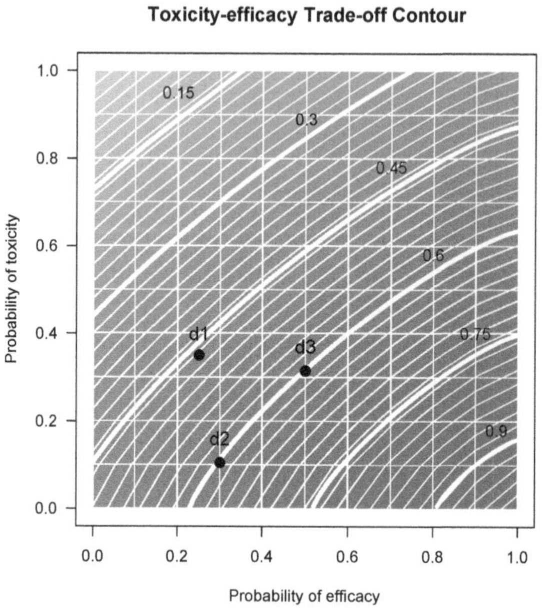

Toxicity-efficacy Trade-off Contour

FIGURE 8.2: Toxicity–efficacy trade-off based on the utility function (8.1), where the lines represent trade-off contours, and the numbers on the contours indicate the desirability of the contours. The dose d_2 is more desirable than d_1 and equally desirable as d_3.

8.2.2   Joint probability model for efficacy and toxicity

The EffTox design uses the standardized dose x_j in its efficacy-toxicity model, defined as

$$x_j = \log(d_j) - \frac{\sum_{j'=1}^{J} \log(d_{j'})}{J}.$$

For example, a trial of 4 doses, $(d_1, d_2, d_3, d_4) = (200, 300, 400, 600)$ mg, would have $(x_1, x_2, x_3, x_4) = (-0.55, -0.14, 0.14, 0.55)$.

The marginal probabilities of efficacy and toxicity at dose d_j are modeled using a logistic model,

$$\pi_{qj} = \text{logit}^{-1}\{\eta_{qj}\}, \quad q = E, T \text{ and } j = 1, \dots, J,$$

where

$$\eta_{Ej} = \mu_E + \beta_{E1} x_j + \beta_{E2} x_j^2$$
$$\eta_{Tj} = \mu_T + \beta_T x_j,$$

with $\beta_T > 0$ to ensure that the marginal toxicity probability π_{Tj} increases with the dose. There is no restriction on β_{E1} and β_{E2}, so that the dose–efficacy curve can take various shapes. To induce correlation between efficacy and toxicity, a Gumbel-Morgenstern copula is used to model the joint distribution of y_E and y_T,

$$f(y_T, y_E, d_j; \boldsymbol{\theta}) = \pi_{Ej}^{y_E}\{1 - \pi_{Ej}\}^{1-y_E} \pi_{Tj}^{y_T}\{1 - \pi_{Tj}\}^{1-y_T} +$$

$$+(-1)^{y_T+y_E}\pi_{Ej}\{1 - \pi_{Ej}\}\pi_{Tj}\{1 - \pi_{Tj}\}\left(\frac{e^\psi - 1}{e^\psi + 1}\right),$$

where $\boldsymbol{\theta} = (\mu_E, \beta_{E1}, \beta_{E2}, \mu_T, \beta_T, \psi)$ is the vector of unknown parameters, and ψ is a real-valued association parameter. Denoting the data of the first n patients in the trial by $D_n = \{y_{Ti}, y_{Ei}, d_{[i]}\}_{i=1}^n$ with $d_{[i]} \in \{d_1, \ldots, d_J\}$ indicating the dose that patient i has received, the joint likelihood is given by

$$L(D_n \mid \boldsymbol{\theta}) = \prod_{i=1}^n f(y_{Ti}, y_{Ei}, d_{[i]}; \boldsymbol{\theta}).$$

Priors on the model parameters are assumed to be normally distributed, with hyperparameter means determined from elicited means of π_{qj} for each (q, j) combination and hyperparameter variances calibrated to obtain a given specified prior effective sample size, $q = E, T$ and $j = 1, \ldots, J$. Additional details are given in Thall and Cook (2004).

8.2.3 Admissible rules

Let ϕ_T be an upper limit of π_{Tj} and ϕ_E be a lower limit of π_{Ej} for a dose to be acceptable, respectively. Values of ϕ_E and ϕ_T should be determined by the clinical investigators based on the trial under consideration, and often vary from one trial to another. For example, $\phi_T = 0.3$ and $\phi_E = 0.2$ means that a dose is regarded as acceptable only when its toxicity probability ≤ 0.3 and efficacy probability ≥ 0.2.

Formally, a dose d_j is defined as acceptable (or admissible) when it satisfies the following two criteria:

$$\text{(Safety criterion)} \quad \Pr(\pi_{Tj} > \phi_T \mid D_n) \leq c_T,$$

$$\text{(Efficacy criterion)} \quad \Pr(\pi_{Ej} < \phi_E \mid D_n) \leq c_E,$$

where c_E and c_T are a large probability threshold such as 0.95. Let $\mathcal{A}(D_n)$ denote the set of admissible doses that satisfy the above safety and efficacy requirements. During the trial, $\mathcal{A}(D_n)$ should be continually updated based on D_n to guide dose escalation and de-escalation. Only doses in $\mathcal{A}(D_n)$ can be used to treat patients.

8.2.4 Dose-finding algorithm

The dose-finding algorithm of the EffTox design, including the adaptive dose assignment, stopping, and selection rules, is given as follows:

(1) The first cohort of patients are treated at the lowest dose, or the starting dose specified by the clinicians.

(2) For each subsequent cohort after the first cohort, we fit the dose-toxicity and dose-efficacy models based on the observed data D_n, and obtain the admissible dose set $\tilde{\mathcal{A}}(D_n) = \mathcal{A}(D_n) \cup \{d_j\}$, where d_j is the lowest untried dose that has acceptable toxicity. If there is no untried dose or the untried doses are all overly toxic, then $\tilde{\mathcal{A}}(D_n) = \mathcal{A}(D_n)$.

(3) If $\tilde{\mathcal{A}}(D_n)$ is empty, the trial should be terminated and no dose is selected; otherwise, the next cohort is treated at the most desirable dose $d_j \in \tilde{\mathcal{A}}(D_n)$ that has the largest estimate of the efficacy-toxicity trade-off $\psi(\pi_{Ej}, \pi_{Tj})$, subject to the constraint that no untried dose may be skipped when escalating. Repeat Steps (2) and (3) if the trial is not terminated.

(4) Stop the trial when the maximum sample size N is reached. If $\mathcal{A}(D_N)$ is not empty, select the OBD as the dose $d_{j*} \in \mathcal{A}(D_N)$ that maximizes the efficacy-toxicity trade-off $\psi(\pi_{Ej*}, \pi_{Tj*})$.

Compared to toxicity-only dose-finding designs such as CRM, by considering the efficacy-toxicity trade-off, the EffTox design better reflects the risk-benefit trade-off underlying medical decisions in practice and identifies the OBD rather than MTD. In addition, the EffTox design uses more data (both toxicity and efficacy data) to make adaptive decisions, and thus is more efficient. The challenges are that the EffTox design is complicated to implement due to complex model (including prior) specification and estimation. In addition, the EffTox design appears sensitive to model misspecification (see Section 8.3.5 and 8.4.4). Although we only cover the EffTox design here, the other model-based phase I–II designs have similar structures as the EffTox design, and share similar challenges.

In the following sections, we introduce several model-assisted phase I-II designs that are easy to implement and that yield performance comparable or better than model-based designs such as EffTox. To use the model-assisted designs, no complicated model fitting is needed, and clinicians can simply look up the decision table to determine the dose for treating the next cohort of patients. In addition, these designs do not assume any parametric model on the dose-toxicity and dose-efficacy curves, and thus are not sensitive to model misspecification.

8.3 U-BOIN design

U-BOIN extends the BOIN design to identify OBD for phase I-II trials (Zhou et al., 2019). It considers a binary toxicity endpoint y_T and a binary efficacy endpoint y_E, where y_E can be tumor response or PD biomarkers measuring biological activities of the drug (e.g., inhibition of an oncogene pathway, gene expression, or proliferation of T cells).

8.3.1 Utility-based risk-benefit trade-off

U-BOIN uses utility to quantify the risk-benefit trade-off, based on possible outcomes that a patient can have. Specifically, given binary y_T and y_E, there are four possible outcomes for any patient in the trial: $(y_T, y_E) = (0, 1) =$ (no toxicity, efficacy); $(0, 0) =$ (no toxicity, no efficacy); $(1, 1) =$ (toxicity, efficacy); and $(1, 0) =$ (toxicity, no efficacy) (Table 8.1). The joint outcome (no toxicity, efficacy) is the most desirable, whereas (toxicity, no efficacy) is the least desirable, and the other two are in between. We quantify the desirability of each outcome by assigning it a utility (score), which should be elicited from physicians to reflect the risk-benefit trade-off underlying their medical decisions.

One way to elicit the utility is given as follows: fix the score of the most desirable outcome (no toxicity, efficacy) as $v_{01} = 100$ and the score of the least desirable outcome (toxicity, no efficacy) as $v_{10} = 0$, and then ask clinicians to use them as a reference to specify scores v_{00} and v_{11} for the other two outcomes, i.e., (no toxicity, no efficacy) and (toxicity, efficacy), respectively to quantify the risk-benefit trade-off under each outcome. v_{00} and v_{11} should be non-negative and $\in [0, 100]$. Table 8.1 lists an example of an elicited utility table. The specification of v_{00} and v_{11} needs to reflect the clinical desirability of the corresponding outcomes: setting $v_{11} \geq v_{00}$ means that having a response is more important than not having a toxicity, and patients are willing to tolerate toxicity in exchange for efficacy; otherwise, $v_{11} < v_{00}$ should be used. From the dose-finding perspective, setting $v_{11} \geq v_{00}$ is usually more meaningful and leads to a more thorough exploration of the dose space than setting $v_{11} < v_{00}$.

Let $(p_{01}, p_{00}, p_{11}, p_{10})$ denote the probabilities of observing the four possible toxicity–efficacy outcomes at a certain dose d. The probability vector $(p_{01}, p_{00}, p_{11}, p_{10})$ typically varies across doses, i.e., as a function of d. For ease of exposition, we suppress the argument d. Averaging over the four possible outcomes, the desirability (or mean utility) of dose d is

$$u = p_{01}v_{01} + p_{00}v_{00} + p_{11}v_{11} + p_{10}v_{10}. \tag{8.2}$$

TABLE 8.1: Utility table for binary toxicity and efficacy endpoints y_T and y_E.

Toxicity (y_T)	Efficacy (y_E)	
	Yes $(= 1)$	No $(= 0)$
No $(= 0)$	$v_{01} = 100$	$v_{00} = 40$
Yes $(= 1)$	$v_{11} = 60$	$v_{10} = 0$

A higher value of u indicates a higher dose desirability in terms of the risk-benefit trade-off.

Revisit the example in Section 8.1 and assume the utility in Table 8.1, when the true toxicity and efficacy probabilities are $(\pi_{T1}, \ldots, \pi_{T5}) = (0.05, 0.12, 0.27, 0.35, 0.50)$ and $(\pi_{E1}, \ldots, \pi_{E5}) = (0.20, 0.35, 0.36, 0.37, 0.38)$, respectively, the desirability of d_3 is $u_3 = 50.8$, and the desirability of d_2 is $u_2 = 56.2$. By using the utility approach, we naturally account for the risk–benefit trade-off and correctly identify that d_2 is more desirable than d_3. Similarly, if $(\pi_{E1}, \ldots, \pi_{E5}) = (0.01, 0.05, 0.30, 0.60, 0.60)$, the desirability of d_4 is $u_4 = 62.0$, which is higher than $u_3 = 47.2$ of dose d_3.

In our experience, clinicians can quickly understand what the utility means and provide utility scores, since the framework resembles clinical practice and the utility scores reflect actual clinical judgment. After completing this process, simulation should be performed to verify the operating characteristics of the design. In some cases, the simulation results may motivate slight modification of some of the numerical utility values.

One possible criticism for using the utility values is that they require subjective input. However, we are inclined to view this as a strength rather than a weakness. The process of specifying the utility requires clinicians to carefully consider the potential risks and benefits of the treatment that underlie their clinical decision making in a more formal way and incorporate that into the trial. In addition, our simulation study and previous studies (Guo and Yuan, 2017; Liu et al., 2018; Murray et al., 2018; Zhou et al., 2019; Lin et al., 2020b) show that the design is generally not sensitive to small differences in the numerical values of the utility as long as the values reflect a similar trend.

Compared to the toxicity-efficacy trade-off contour approach used by the EffTox assign, the utility approach has several advantages. First, it is simpler and only requires elicitation of the utility score of element outcomes that have transparent clinical interpretation. In contrast, the process of constructing the toxicity-efficacy trade-off contour is complicated. More problematically, it assumes that the contour must follow a specific function form, equation (8.1), which lacks clinical interpretation and justification, and is likely to be false. In our experience, it is often much easier for clinicians to understand and quantify the risk-benefit trade-off based on patient outcomes than based on toxicity and efficacy probabilities.

Second, the utility approach is highly flexible and includes other risk-benefit trade-off methods as special cases (Zhou et al., 2019; Lin et al., 2020b). For example, by imposing the constraint $v_{00} + v_{11} = 100$, we have

$$
\begin{aligned}
u &= 100p_{01} + v_{00}p_{00} + v_{11}p_{11} + 0p_{10} \\
&= (v_{00} + v_{11})p_{01} + v_{00}p_{00} + v_{11}p_{11} \\
&= v_{00}(p_{01} + p_{00}) + v_{11}(p_{01} + p_{11}) \\
&= v_{11}\left\{ \pi_E + \frac{v_{00}}{v_{11}}(1 - \pi_T) \right\},
\end{aligned}
$$

where $\pi_T = p_{11} + p_{10}$ and $\pi_E = p_{01} + p_{11}$ are the marginal probabilities of toxicity and efficacy at dose d, respectively. Therefore, the risk-benefit trade-off approach based on the marginal toxicity and efficacy probabilities,

$$u' = \pi_E - w\pi_T, \tag{8.3}$$

is a special case of (8.2) with $w = v_{00}/v_{11}$ and $v_{00} + v_{11} = 100$. For example, the utility shown in Table 8.1 satisfies $v_{00} + v_{11} = 100$, and thus it is equivalent to the simplified trade-off

$$u' = \pi_E - 2/3\pi_T. \tag{8.4}$$

This toxicity–efficacy trade-off means that a 1% toxicity increase in conjuncture with a w% increase of efficacy does not change the desirability of the treatment. In other words, the weight $w = v_{00}/v_{11}$ implicitly determines the relative importance between toxicity and efficacy.

Another important special case of the utility approach is to set $v_{00} = 0$ and $v_{11} = 100$ to turn off the toxicity-efficacy trade-off. Under this utility, we favor the dose with the highest efficacy, which is appropriate for trials aiming to identify the dose yielding the highest efficacy. The safety of the dose is safeguarded by the admissible rules described below.

Third, the utility approach is highly scalable. It is directly applicable to categorical y_T and y_E with more than two levels. Table 8.2 provides an example of the utility approach for a trinary toxicity endpoint (minor/moderate/severe) and a trinary efficacy endpoint (progressive disease (PD)/stable disease (SD)/partial response or complete response (PR/CR)). In addition, Lin et al. (2020b) shows that it is similarly straightforward to accommodate more than two endpoints. In contrast, it is not clear how to apply the toxicity-efficacy trade-off contour approach to these cases.

8.3.2 Statistical model

Let $D = (y_{01}, y_{00}, y_{11}, y_{10})$ denote the observed data at dose d, where $(y_{01}, y_{00}, y_{11}, y_{10})$ denote the numbers of patients having outcomes (no toxicity, efficacy), (no toxicity, no efficacy), (toxicity, efficacy), and (toxicity, no

TABLE 8.2: Utility table for trinary efficacy and toxicity endpoints

Toxicity	Efficacy		
	CR/PR	SD	PD
Minor	100	60	35
Moderate	65	30	25
Severe	30	15	0

PD: progressive disease, SD: stable disease, PR: partial response, CR: complete response.

efficacy), respectively. Here, $n = y_{01} + y_{00} + y_{11} + y_{10}$ is the number of patients treated at dose d. The numbers of patients who have experienced toxicity and efficacy are $n_T = y_{11} + y_{10}$ and $n_E = y_{01} + y_{11}$, respectively. Subscript j (i.e., dose level) is suppressed in this and next sections for notational brevity.

Under the Bayesian paradigm, the U-BOIN design assumes a Dirichlet-multinomial model,

$$
\begin{aligned}
(y_{01}, y_{00}, y_{11}, y_{10}) &\sim \text{Multinomial}(n; p_{01}, p_{00}, p_{11}, p_{10}) \\
(p_{01}, p_{00}, p_{11}, p_{10}) &\sim \text{Dirichlet}(\alpha_{01}, \alpha_{00}, \alpha_{11}, \alpha_{10})
\end{aligned}
$$

where $(\alpha_{01}, \alpha_{00}, \alpha_{11}, \alpha_{10})$ are hyperparameters, representing the prior numbers of events for four outcomes, with the total prior sample size $n_0 = \alpha_{01} + \alpha_{00} + \alpha_{11} + \alpha_{10}$. The posterior distribution of $(p_{01}, p_{00}, p_{11}, p_{10})$ based on D is

$$
(p_{01}, p_{00}, p_{11}, p_{10}) \mid D \sim \text{Dirichlet}(\alpha_{01} + y_{01}, \alpha_{00} + y_{00}, \alpha_{11} + y_{11}, \alpha_{10} + y_{10}).
$$

Therefore, the posterior mean estimate of u is given by

$$
u = \hat{p}_{01} v_{01} + \hat{p}_{00} v_{00} + \hat{p}_{11} v_{11} + \hat{p}_{10} v_{10}, \tag{8.5}
$$

where $\hat{p}_{ab} = (\alpha_{ab} + y_{ab})/(n + n_0)$ is the posterior mean estimate of p_{ab}, $a, b \in (0, 1)$.

It follows that the marginal posterior distributions of π_T and π_E follow the beta distribution,

$$
\begin{aligned}
\pi_T \mid D &\sim \text{Beta}(\alpha_T + n_T, \beta_T + n - n_T), \\
\pi_E \mid D &\sim \text{Beta}(\alpha_E + n_E, \beta_E + n - n_E),
\end{aligned} \tag{8.6}
$$

where the hyperparameters $\alpha_T = \alpha_{11} + \alpha_{10}, \alpha_E = \alpha_{11} + \alpha_{01}, \beta_T = \alpha_{01} + \alpha_{00}, \beta_E = \alpha_{10} + \alpha_{00}$. For example, specifying $\alpha_T = \alpha_E = \beta_T = \beta_E = 1$ (e.g., $\alpha_{01} = \alpha_{00} = \alpha_{11} = \alpha_{10} = 0.5$) leads to $\text{Unif}(0, 1)$ prior distributions for π_T and π_E.

8.3.3   Admissible rules

The U-BOIN design uses similar admissible rules as those of the EffTox design. Let ϕ_T and ϕ_E be the clinician-specified toxicity upper limit and efficacy lower limit, respectively. Generally, ϕ_E can take the value of the target response rate specified for a standard phase II trial. Because U-BOIN considers the toxicity–efficacy trade-off, the value of ϕ_T should be set slightly higher (e.g, 0.05) than the target toxicity rate used in conventional toxicity-based phase I designs. For example, if 25% is an appropriate target toxicity rate that the conventional phase I design used, then $\phi_T = 30\%$ is a reasonable choice for U-BOIN.

Given the observed interim data D, a dose d_j is defined as acceptable (or admissible) when it satisfies both of the following two criteria:

$$
\text{(Safety criterion)} \quad \Pr(\pi_{Tj} > \phi_T \mid D) \le c_T, \tag{8.7}
$$

$$
\text{(Efficacy criterion)} \quad \Pr(\pi_{Ej} < \phi_E \mid D) \le c_E, \tag{8.8}
$$

where c_E and c_T is a probability threshold. In general, we recommend $c_T = 0.95$ and $c_E = 0.90$ as the default, which can be calibrated by simulation. These cutoffs seem high, but actually are appropriate as their purpose is to rule out excessively toxic and ineffective doses. Among admissible doses, the dose assignment rule will allocate patients to the most desirable dose. In other words, even if the admissible dose set includes some doses that are not particularly safe or efficacious, the design likely will not assign patients to these suboptimal doses. Due to the large uncertainty of small sample sizes, using small values for c_T and c_E will inadvertently eliminate the doses that are actually admissible and thus affect the operating characteristics of the design.

Evaluation of the admissibility of a dose is straightforward under the U-BOIN design because $f(\pi_{Tj} \mid D)$ and $f(\pi_{Ej} \mid D)$ follow beta distributions given by (8.6). In contrast, evaluation of the admissibility of a dose is much more complicated in the EffTox design and requires fitting the complex Gumbel-Morgenstern copula model using a Markov chain Monte Carlo method. Let $\mathcal{A}(D)$ denote the set of admissible doses. During the trial, only doses in $\mathcal{A}(D)$ can be used to treat patients.

8.3.4 Dose-finding algorithm

The U-BOIN design consists of two seamless, connected stages (Figure 8.3). The objective of Stage I is to quickly explore the dose space to identify a set of admissible doses that are reasonably efficacious and safe for Stage II. The objective of Stage II is to optimize the dose based on the utility. In Stage I, we conduct dose escalation based on the BOIN design using only y_T, but y_E is also collected and will be used for decision making in Stage II. Given the exploratory nature of Stage I, if y_T has more than two categories, we dichotomize it as DLT/no-DLT to facilitate the exploration of the dose space. This is in line with clinical practice and serves well for the purpose of Stage I.

Consider a phase I-II trial with J doses, $d_1 < \cdots < d_J$, let $\hat{\pi}_{Tj} = n_{Tj}/n_j$ be the observed toxicity rate of dose d_j. Based on the BOIN's dose escalation and de-escalation boundaries (λ_e, λ_d), the dose-finding algorithm of U-BOIN in Stage I proceeds as follows:

A1 Patients in the first cohort are treated at the lowest dose, or a physician-specified dose.

A2 Suppose j is the current dose level. To assign a dose to the next cohort of patients:

- If $\hat{\pi}_{Tj} \leq \lambda_e$, escalate the dose to level $j + 1$.

- If $\hat{\pi}_{Tj} > \lambda_d$, de-escalate the dose to level $j - 1$.

- Otherwise, i.e., $\lambda_e < \hat{\pi}_{Tj} \leq \lambda_d$, stay at the current dose level j.

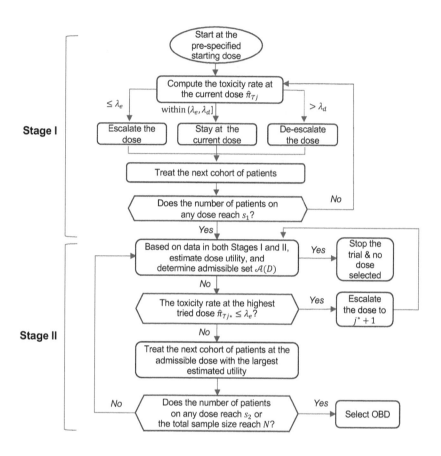

FIGURE 8.3: Diagram of the utility-based Bayesian optimal interval (U-BOIN) design with the pick-the-winner strategy for patient allocation in stage II.

A3   Repeat Step A2 until the number of patients treated on one of the doses reaches s_1, and then move to Stage II. Zhou et al. (2019) recommended $s_1 = 12$ as the default value, while $s_1 = 9$ to 15 generally yields good operating characteristics.

Stage II proceeds as follows:

B1   Let j^* denote the highest dose level that has been tried. If $\hat{\pi}_{Tj^*} \leq \lambda_e$ and j^* is not the highest dose in the trial, escalate the dose to level $j^* + 1$ for treating the next cohort of patients; otherwise, proceed to Step B2.

B2 Update the admissible dose set $\mathcal{A}(D)$ based on the interim data D. If no dose is admissible, terminate the trial and no dose should be selected as OBD. Otherwise, use one of the following strategies to allocate the next cohort of patients:

   (i) Equally randomize the next cohort of patients to a dose in $\mathcal{A}(D)$.

   (ii) Assign the next cohort of patients to the admissible dose that has the largest posterior mean utility $\hat{u}$.

   (iii) Adaptively randomize the next cohort of patients to a dose in $\mathcal{A}(D)$ with a probability proportional to its $\hat{u}$.

B3 Repeat Steps B1 and B2 until the maximum sample size N is exhausted or the number of patients treated at one of the doses reaches s_2 ($> s_1$), and then select the OBD as the admissible dose (i.e., $\in \mathcal{A}(D)$) that has the largest posterior mean utility.

In Stage I, following the BOIN design, we impose an overdose control rule as follows: if $\Pr(\pi_{Tj} > \phi_T \mid D_j) > 0.95$ and $n_j \geq 3$, dose level j and higher are eliminated from the trial, and the trial is terminated if the lowest dose level is eliminated, where $\Pr(\pi_{Tj} > \phi_T \mid D_j) > 0.95$ is evaluated based a beta-binomial model with the uniform prior. Once the trial moves on to Stage II, this overdose control rule is seamlessly merged as the safety criterion of the inadmissible rule defined in Section 8.3.3.

For Stage II Step B1, the reason that we perform dose escalation when $\hat{\pi}_{Tj^*} \leq \lambda_e$ is to allow the trial to continue exploring the dose space, given that the highest tried dose is safe, to reduce the risk of being stuck at a local suboptimal dose due to large variation caused by small sample size.

In Stage II Step B2, the three strategies generally yield similar performance in identifying the OBD, but have a different emphasis on patient allocation. Determining which one to use should be based on specific trial considerations. Strategy (i) (i.e., equal randomization) is the easiest to implement and allows more uniform learning of admissible doses. In addition, equal randomization does not require efficacy readout, avoiding the common logistic difficulty that the efficacy endpoint may take a relatively long time to be ascertained. Also, equal randomization is useful to balance confounders across admissible doses to facilitate reliable identification of OBD particularly when patients are highly heterogenous (e.g., all-comer trials). Equal randomization can be modified to a fixed-ratio randomization. For example, based on the number of patients allocated at Stage I, we choose a fixed randomization ratio such that at the end of Stage II, the sample size at each admissible dose is expected to be the same.

In comparison, Strategy (ii), i.e., pick the winner, aims to maximize patient benefit by assigning the next cohort to the currently estimated optimal dose. This approach tends to assign more patients to OBD than Strategy (i). As a trade-off, in certain scenarios it may have a slightly lower probability of selecting OBD as it may be stuck at local suboptimal doses, especially when the sample size is small. This issue, however, is generally minor. Due the small sample size, the estimated optimal dose often associates high uncertainty, and varies when data accumulate. Thus, Strategy (ii) actually behaves rather like adaptive randomization, i.e., Strategy (iii), although the latter exibits slightly more randomness by design. Compared to Strategy (iii), Strategy (ii) has similar performance, but is easier to understand and implement as described below. Thus, we generally recommend Strategy (i) or (ii). See Zhou et al. (2019) for more details and the numerical comparison of the three patient allocation strategies.

Compared to the EffTox design, one prominent advantage of U-BOIN is that its dose assignment rules can be pretabulated in decision tables and included in the trial protocol before the trial starts. To conduct the trial, no complicated calculation is needed. The investigator can simply use the decision tables to determine dose assignment. Specifically, Stage I and Stage II Step B1 are guided by the BOIN design decision table (see Section 3.4 for details), and Steps B2 and B3 can be implemented based on the pre-calculated utility table. Table 8.3 provides the utility table when the number of patients at a dose is three or six, with the elicited utility scores in Table 8.1, $\phi_T = 0.3$, $\phi_E = 0.3$, $c_T = 0.95$, and $c_E = 0.9$. A similar utility table can be generated for any given n using the U-BOIN software described later. To illustrate the use of the table, suppose at an interim of the trial, at d_1, three patients were treated with no toxicity and one efficacy, and at d_2, six patients were treated with two toxicity and two efficacy. By looking up the table, the estimated utility of d_1 is 57.5 and the estimated utility of d_2 is 47.1. Thus, if we use Strategy (ii) to allocate patients, the next cohort should be treated at dose level one as it has higher desirability.

Of note, in Table 8.3, $\hat{u}$ depends only on n_T and n_E because the utility score used (Table 8.1) satisfies $v_{00} + v_{11} = 100$. In this case, as described previously, u is a function of π_T and π_E, see equation (8.3), thus $\hat{u}$ depends only on n_T and n_E. In the general case that $v_{00} + v_{11} \neq 100$, $\hat{u}$ is a function of y_{10}, y_{00}, y_{01}, and y_{11}. The utility table still can be constructed, but with more entries because of more possible data patterns given n.

8.3.5   Operating characteristics

Zhou et al. (2019) conducted a simulation study to compare the operating characteristics of U-BOIN to the EffTox design. The simulation considered $J = 5$ doses, and the total sample size $N = 54$ patients with $s_1 = 12$ (i.e., Stage I stopping cutoff). The lower limit for efficacy was $\phi_E = 0.2$ and the upper limit for toxicity was $\phi_T = 0.30$. For the admissible rules (i.e., equations (8.7) and (8.8)), probability cutoffs were set as $c_T = 0.95$ and $c_E = 0.9$). Table 8.4 shows a part of the simulation results, see Zhou et al. (2019) for details. In general, U-BOIN outperformed EffTox with higher OBD selection percentages and often allocated more patients to OBD. EffTox performed

TABLE 8.3: Estimated mean utility (i.e., $\hat{u}$) given the number of patients treated at a dose is 3 or 6 with the elicited utility scores in Table 8.1, toxicity upper limit $\phi_T = 0.3$ and efficacy lower limit $\phi_E = 0.3$.

n_T	n_E	$\hat{u}$		n_T	n_E	$\hat{u}$
			$n = 3$			
0	0	42.5		1	3	77.5
0	1	57.5		2	0	22.5
0	2	72.5		2	1	37.5
0	3	87.5		2	2	52.5
1	0	32.5		2	3	67.5
1	1	47.5		> 2	Any	0
1	2	62.5				
			$n = 6$			
0	0	41.4		2	1	38.6
0	< 1	0		2	1	38.6
0	1	50		2	2	47.1
0	2	58.6		2	3	55.7
0	3	67.1		2	4	64.3
0	4	75.7		2	5	72.9
0	5	84.3		2	6	81.4
0	6	92.9		3	< 1	0
1	< 1	0		3	1	32.9
1	1	44.3		3	2	41.4
1	2	52.9		3	3	50
1	3	61.4		3	4	58.6
1	4	70		3	5	67.1
1	5	78.6		3	6	75.7
1	6	87.1		> 3	Any	0
2	< 1	0				

Note: n is the number of the patients at a specific dose, n_T and n_E are respectively the numbers of toxicity and efficacy at that dose. "0" denotes that the dose should be eliminated as not admissible because of high toxicity or low efficacy.

well in scenario 2, but (relatively) poorly in scenarios 1 and 3, showing its sensitivity to model misspecification, i.e., how well the true dose-toxicity or dose-efficacy relationship can be approximated by the model assumed.

8.4 BOIN12 design

BOIN12 is another efficient model-assisted design to find OBD. BOIN12 takes the single-stage approach, which uses toxicity and efficacy jointly throughout the trial to determine dose escalation/de-escalation. This differs from

TABLE 8.4: Comparison of U-BOIN with EffTox, including the selection percentage (Sel %) and the average number of patients treated at each dose (No. of pts). The optimal biological dose (OBD) is bolded.

Design		Dose Level				
		1	2	3	4	5
		Scenario 1				
	π_T	0.02	**0.15**	0.30	0.45	0.60
	π_E	0.20	**0.65**	0.65	0.65	0.65
	u	43.0	**69.0**	63.0	56.0	50.0
EffTox	Sel %	2.0	**50.0**	45.0	2.0	0.0
	No. of pts	4.3	**22.7**	23.8	2.8	0.4
U-BOIN	Sel %	1.7	**72.9**	22.4	2.8	0.0
	No. of pts	6.2	**29.9**	13.8	3.5	0.5
		Scenario 2				
	π_T	0.03	0.08	**0.15**	0.28	0.40
	π_E	0.10	0.22	**0.60**	0.60	0.60
	u	36.0	43.0	**66.0**	60.0	55.0
EffTox	Sel %	0.0	4.0	**60.0**	29.0	7.0
	No. of pts	3.4	4.9	**26.4**	14.2	5.1
U-BOIN	Sel %	1.1	3.2	**65.7**	24.9	4.3
	No. of pts	4.9	7.5	**24.4**	12.7	4.3
		Scenario 3				
	π_T	0.05	0.07	0.10	0.12	**0.16**
	π_E	0.35	0.45	0.50	0.55	**0.75**
	u	53.0	59.0	61.0	64.0	**75.0**
EffTox	Sel %	10.0	12.0	26.0	24.0	**29.0**
	No. of pts	8.6	7.9	14.2	11.1	**12.1**
U-BOIN	Sel %	5.9	11.7	13.1	13.6	**55.7**
	No. of pts	7.0	8.8	9.0	9.1	**20.1**

Note: π_T and π_E are the toxicity probability and efficacy probability, respectively. u is the mean utility or desirability.

U-BOIN's two-stage approach, which first performs toxicity-based dose escalation, and then switches to dose optimization jointly based on toxicity and efficacy. As a result, U-BOIN and BOIN12 demonstrate different design characteristics and are suitable for different trial objectives.

U-BOIN is more appropriate for the case where both identification of OBD and evaluation of the dose-toxicity profile of the drug are of interest, whereas BOIN12 is a better choice when the primary goal is to identify OBD. For the purpose of finding the OBD, BOIN12 is more efficient and often requires a smaller sample size than U-BOIN. For example, for drugs with a low, flat dose-toxicity curve and an increase-then-plateau dose-efficacy curve, U-BOIN will first escalate all the way up to the highest dose, and then turn around to search for OBD. In contrast, BOIN12 will quickly reach and stay at OBD because it considers both efficacy and toxicity from the beginning of the trial. Compared to BOIN12, U-BOIN has some operational advantages. BOIN12 requires that the efficacy endpoint must be scored quickly enough so that the dose assignment decision is timely for each new cohort. U-BOIN is lenient on this requirement as its first stage often "buys" sufficient time to evaluate the efficacy endpoint, especially when the equal or fixed-ratio randomization is employed in Stage II of U-BOIN. Section 8.5 introduces an extension of BOIN12 that is able to handle delayed efficacy or/and toxicity endpoints.

8.4.1   Utility estimation using quasi-binomial likelihood

Similar to U-BOIN, BOIN12 uses utility to quantify the risk-benefit trade-off and desirability of each dose. As described in Section 8.3.1, the utility approach has the advantage of being easy to understand and implement, and it is highly flexible and scalable to accommodate various types of endpoints. We adopt the same notation as the previous section and focus on a binary toxicity endpoint y_T and binary efficacy endpoint y_E.

Let $(p_{01}, p_{00}, p_{11}, p_{10})$ denote the probabilities of observing the four possible values of (y_T, y_E), i.e., $(0, 1)$, $(0, 0)$, $(1, 1)$, and $(1, 0)$, at a certain dose d. For notational brevity, we suppress that $(p_{01}, p_{00}, p_{11}, p_{10})$ is a function of d. Let $(v_{01}, v_{00}, v_{11}, v_{10})$ denote the utility scores, which are elicited from clinicians to quantify the desirability of the four possible outcomes of (y_T, y_E), see Table 8.1 for an example. Similar to Section 8.3.1, we set the utility of the most desirable outcome $v_{01} = 100$ to fix the scale of the score. The desirability (or mean utility) of dose d is given by

$$u = p_{01}v_{01} + p_{00}v_{00} + p_{11}v_{11} + p_{10}v_{10}. \tag{8.9}$$

A higher value of u indicates a higher dose desirability in terms of the risk-benefit trade-off.

The BOIN12 design makes decisions of dose escalation/de-escalation based on $f(u \mid D)$, the posterior distribution of u given the interim data D. A straightforward approach is to derive the posterior $f(u \mid D)$ based on the definition (8.9) and the Dirichlet-multinomial model, described in Section 8.3.2.

Lin et al. (2020b) proposed a more straightforward approach by directly modeling u based on the quasi-binomial likelihood theory (Papke and Wooldridge, 1996; Yuan et al., 2007). Define the standardized desirability $\tilde{u} = u/100$. Because $\tilde{u} \in [0, 1]$ and takes a form of the weighted average of $(p_{01}, p_{00}, p_{11}, p_{10})$, it can be viewed as a probability and modeled using the binomial distribution with "quasi-binomial" data (x, n), where

$$x = \frac{100y_{01} + v_{00}y_{00} + v_{11}y_{11}}{100}. \tag{8.10}$$

This quasi-number x can be interpreted as the number of "events" observed from n patients treated at dose d, given the event probability $\tilde{u}$. This interpretation follows because $E(x) = n \times \tilde{u}$. Unlike the standard binomial data, x is a real value that can take any value between 0 and n instead of integers only, and thus it is known as quasi-binomial. When $v_{00} + v_{11} = 100$, the expression of x can be further simplified to

$$x = \frac{v_{11} n_E + v_{00}(n - n_T)}{100}.$$

The quasi-binomial likelihood of the unknown $\tilde{u}$ based on the interim data D is

$$L(D \mid \tilde{u}) \propto \tilde{u}^x \left(1 - \tilde{u}\right)^{n-x}.$$

Under the Bayesian framework, assigning $\tilde{u}$ a Beta prior, i.e., $\tilde{u} \sim$ Beta(α_u, β_u), the posterior distribution of $\tilde{u}$ arises as

$$\tilde{u} \mid D \sim \text{Beta}(\alpha_u + x, \beta_u + n - x). \tag{8.11}$$

By default, the non-informative uniform prior distribution with $\alpha_u = \beta_u = 1$ is used. A similar quasi-binomial approach was used by Yuan et al. (2007) to model the normalized equivalent toxicity score that maps toxicity grades into a fraction of dose limiting toxicity under the parametric CRM model.

8.4.2   Admissible rules and dose comparison

BOIN12 uses the same admissible rules as those of U-BOIN, described in Section 8.3.3. Briefly, let ϕ_T and ϕ_E denote the toxicity upper limit and efficacy lower limit, respectively, specified by clinicians. Generally, ϕ_E may take the value of the target response rate specified for a standard phase II trial. The value of ϕ_T should be set slightly higher (e.g., 0.05) than the target toxicity rate used in conventional toxicity-based phase I designs to provide additional space for toxicity–efficacy trade-off. For example, if 30% is an appropriate target toxicity rate that the conventional phase I design used, then $\phi_T = 35\%$ is a reasonable choice for BOIN12.

Given the interim data D, a dose d_j is defined as acceptable (or admissible) when it satisfies both of the following two criteria:

$$\text{(Safety criterion)} \quad \Pr(\pi_{Tj} > \phi_T \mid D) \le c_T,$$

$$\text{(Efficacy criterion)} \quad \Pr(\pi_{Ej} < \phi_E \mid D) \le c_E,$$

where c_E and c_T are probability thresholds. These two posterior quantities can be evaluated based on the posterior of π_T and π_E, given by equation (8.6). Denote the set of admissible doses by $\mathcal{A}(D)$, which should be updated with interim data D. Only doses in $\mathcal{A}(D)$ can be used to treat patients. In general, we recommend $c_T = 0.95$ and $c_E = 0.90$ as the default, which can be further calibrated by simulation, see Section 8.3.3 for more discussion on the choice of c_T and c_E.

In BOIN12, the posterior probability $\text{PP}_j = \Pr(u_j > u_b \mid D_j)$ is used to determine the most desirable dose within $\mathcal{A}(D)$ for treating a new cohort of patients, where D_j is the data observed at d_j. Here, u_b is a prespecified utility benchmark used to evaluate dose desirability. Between two admissible doses d_j and $d_{j'}$,

- If $\text{PP}_j > \text{PP}_{j'}$, then d_j is more desirable than $d_{j'}$.

- If $\text{PP}_j < \text{PP}_{j'}$, then d_j is less desirable than $d_{j'}$.

- If $\text{PP}_j = \text{PP}_{j'}$, then d_j and $d_{j'}$ are equally desirable.

In principle, the third case occurs if and only if $D_j = D_{j'}$. The posterior probability PP_j automatically accounts for the uncertainty in estimating u, leading to more reasonable decision making of dose escalation and de-escalation. In contrast, most existing designs use the point estimate $\hat{u} = x/n$ to choose the optimal dose, which ignores the uncertainty of the estimate. To see the importance of accounting for estimation uncertainty, suppose dose d_j has been used to treat two patients and $\hat{u}_j = 60$, and dose $d_{j'}$ has been used to treat 10 patients and $\hat{u}_{j'} = 59$. Although $\hat{u}_j > \hat{u}_{j'}$, because $\hat{u}_{j'}$ is much more reliable, a better decision actually is to choose $d_{j'}$ for treating the next cohort of patients as stipulated by the posterior probability approach.

Evaluating PP_j requires specifying a benchmark u_b for comparison. The value of u_b can be elicited by clinicians to reflect their expectation. Lin et al. (2020b) recommended the following default value of u_b that yields desirable operating characteristics in a variety of scenarios. Given ϕ_T and ϕ_E and assuming independence between toxicity and efficacy, the highest utility that is deemed undesirable is given by

$$\underline{u} = 100\phi_E(1 - \phi_T) + v_{00}(1 - \phi_T)(1 - \phi_E) + v_{11}\phi_T\phi_E.$$

The recommended default value of u_b is

$$u_b = (100 + \underline{u})/2,$$

representing that a utility value lies in the middle between $\underline{u}$ and the maximum utility 100. Alternatively, one can also use the weighted average as the choice for u_b, i.e., $u_b = \omega 100 + (1 - \omega)\underline{u}$, where $\omega \in [0, 1]$ is a non-negative weight. In general, the weight ω controls the aggressiveness of the trial: the larger the value of ω, the more aggressive the dose exploration.

One prominent feature of BOIN12 is that the desirability can be pre-calculated and included in the trial protocol, due to the use of the quasi-binomial likelihood. Specifically, given the maximum sample size N, all possible outcome combinations (n, n_T, n_E) (or $(y_{01}, y_{00}, y_{11}, y_{10})$) can be enumerated, and the corresponding posterior probabilities PP_j can be computed based on (8.11). By sorting all the possible values of PP_j from the smallest to the largest, we can assign the ordered value from $0, 1, \dots$ to each possible outcome combination. These ordered values are called rank-based desirability scores (RDS). RDS allows for a simpler presentation of the desirability of a dose using integers. Comparing the posterior probabilities PP_j between two doses is equivalent to comparing their corresponding RDS; therefore, RDS can be used as a convenient way to identify a more desirable dose. For example, based on the utility in Table 8.1, Table 8.5 lists RDS with a cohort size of three to nine patients treated at a dose, with the highest acceptable toxicity probability $\phi_T = 0.35$ and the lowest acceptable efficacy probability $\phi_E = 0.25$.

During the trial, the desirability of a dose can be simply determined as follows: count the number of patients treated at that dose n, the number of patients who experienced toxicity n_T, and the number of patients who experienced efficacy n_E (and the number of patients who experienced efficacy without toxicity when $v_{00} + v_{11} \neq 100$), then look up Table 8.5 to determine its RDS.

For example, suppose at a certain point of the trial, the numbers of patients treated at the first three doses are three, six, and three; the numbers of toxicities are zero, one, and two; and the numbers of efficacy outcomes are zero, three, and one. By looking up Table 8.5, the RDS of the first three doses are 35, 56, and 31, respectively. Because dose d_2 has the highest RDS, it is more desirable than d_1 and d_3. Table 8.5 assumes a cohort size of three, but the BOIN12 design is flexible and can allow any prespecified cohort size. The RDS table can be easily generated using the BOIN12 companion software.

It is worth noting the difference between BOIN12 and U-BOIN. U-BOIN uses the posterior mean of u (i.e., $\hat{u}$) to determine the most desirable dose within $\mathcal{A}(D)$ for treating a new cohort of patients. The utility table of U-BOIN (e.g., Table 8.3) lists the value of $\hat{u}$ for each possible interim data, not the rank of $\hat{u}$. Nevertheless, in principle, for U-BOIN, a RDS table similar to that of BOIN12 can also be developed by replacing the value of $\hat{u}$ with its rank among all possible values.

8.4.3   Dose-finding rules

Suppose that J dose levels are investigated, $d_1 < \cdots < d_J$. We use subscript j to indicate variables associated with dose d_j. Let $\hat{\pi}_{Tj} = n_{Tj}/n_j$ denote the observed toxicity rate at dose d_j, and (λ_e, λ_d) denote the dose escalation and de-escalation boundaries from the standard BOIN design. The dose-finding rule of the BOIN12 design is illustrated in Figure 8.4 and described as follows:

1. Patients in the first cohort are treated at the lowest dose d_1, or the physician-specified dose.

2. Based on the interim data D, we obtain RDS for each dose using the RDS table (e.g., Table 8.5). Suppose j is the current dose level. To assign a dose to the next cohort of patients,

TABLE 8.5: Rank-based desirability score (RDS) table for the BOIN12 design.

n	n_T	n_E	RDS	n	n_T	n_E	RDS	n	n_T	n_E	RDS
0	0	0	60	6	3	2	22	9	2	4	45
3	0	0	35	6	3	3	38	9	2	5	58
3	0	1	55	6	3	4	51	9	2	6	70
3	0	2	76	6	3	5	67	9	2	7	83
3	0	3	91	6	3	6	81	9	2	8	92
3	1	0	24	6	4	0	1	9	2	9	98
3	1	1	44	6	4	1	6	9	3	0	E
3	1	2	63	6	4	2	15	9	3	1	7
3	1	3	80	6	4	3	27	9	3	2	14
3	2	0	13	6	4	4	42	9	3	3	25
3	2	1	31	6	4	5	56	9	3	4	36
3	2	2	48	6	4	6	72	9	3	5	49
3	2	3	69	6	≥ 5	Any	E	9	3	6	61
3	≥ 3	Any	E	9	0	0	E	9	3	7	74
6	0	0	22	9	0	1	25	9	3	8	85
6	0	1	38	9	0	2	36	9	3	9	94
6	0	2	51	9	0	3	49	9	4	0	E
6	0	3	67	9	0	4	61	9	4	1	3
6	0	4	81	9	0	5	74	9	4	2	9
6	0	5	93	9	0	6	85	9	4	3	17
6	0	6	100	9	0	7	94	9	4	4	29
6	1	0	15	9	0	8	99	9	4	5	40
6	1	1	27	9	0	9	102	9	4	6	53
6	1	2	42	9	1	0	E	9	4	7	65
6	1	3	56	9	1	1	17	9	4	8	78
6	1	4	72	9	1	2	29	9	4	9	88
6	1	5	87	9	1	3	40	9	5	0	E
6	1	6	96	9	1	4	53	9	5	1	2
6	2	0	8	9	1	5	65	9	5	2	5
6	2	1	19	9	1	6	78	9	5	3	10
6	2	2	34	9	1	7	88	9	5	4	20
6	2	3	47	9	1	8	97	9	5	5	32
6	2	4	64	9	1	9	101	9	5	6	45
6	2	5	77	9	2	0	E	9	5	7	58
6	2	6	90	9	2	1	10	9	5	8	70
6	3	0	4	9	2	2	20	9	5	9	83
6	3	1	12	9	2	3	32	9	≥ 6	Any	E

Note: n is the number of the patients at a specific dose, n_T and n_E are respectively the numbers of toxicity and efficacy at that dose. "E" denotes that the dose should be eliminated because it does not satisfy the safety and efficacy admissible criteria (i.e., not admissible because of high toxicity or low efficacy).

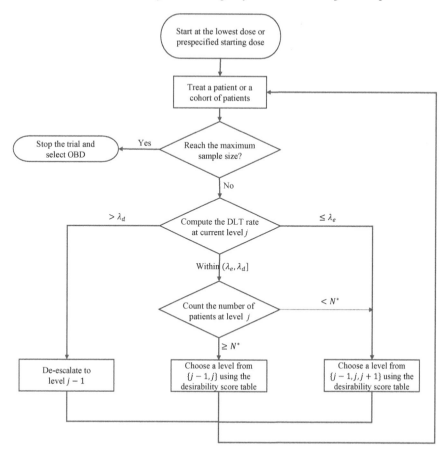

Target toxicity rate ϕ_T	0.20	0.25	0.30	0.35	0.40
Escalation boundary λ_e	0.157	0.197	0.236	0.276	0.316
De-escalation boundary λ_d	0.238	0.298	0.359	0.419	0.480

FIGURE 8.4: The schema of the Bayesian optimal interval phase I–II (BOIN12) design, where (λ_e, λ_d) are a pair of optimized dose escalation and de-escalation boundaries adopted from the BOIN design, and N^* is a pre-specified sample size cutoff (e.g., $N^* = 6$). DLT, dose-limiting toxicity; OBD, optimal biologic dose.

(a) If $\hat{\pi}_{Tj} > \lambda_d$, de-escalate the dose to level $j - 1$.

(b) If $\hat{\pi}_{Tj} > \lambda_e$ and $n_j \geq N^*$, say $N^* = 6$, choose the level from $\{j - 1, j\}$ that has a higher RDS.

(c) Otherwise, if $\lambda_e < \hat{\pi}_{Tj} \leq \lambda_d$ and $n_j < N^*$, or $\hat{\pi}_{Tj} \leq \lambda_e$, choose the level from $\{j - 1, j, j + 1\}$ that has the highest RDS.

3. Repeat Step 2 until the maximum sample size N is reached.

Step 2(a) says that if the observed toxicity rate is high (i.e., $\hat{\pi}_{Tj} > \lambda_d$), we should de-escalate the dose. Step 2(b) says that if there are sufficient data ($n_j \geq N^*$) to support that the toxicity probability of the current dose d_j is moderate (i.e., $\lambda_e < \hat{\pi}_{Tj} \leq \lambda_d$), escalating the dose to d_{j+1} may cause overdosing. We then examine whether the adjacent lower dose d_{j-1} provides a better treatment benefit (i.e., desirability) than the current dose d_j. If so, we treat the next cohort of patients at dose level $j - 1$; otherwise stay at the current level j. Rule 2(c) says that if the observed toxicity rate at the current dose d_j is low ($\hat{\pi}_{Tj} \leq \lambda_e$) or moderate ($\lambda_e < \hat{\pi}_{Tj} \leq \lambda_d$) but with high uncertainty ($n < N^*$), then we examine whether the adjacent higher dose d_{j+1} or the lower dose d_{j-1} can provide a better treatment benefit (i.e., desirability) than the current dose d_j. The next cohort of patients will be treated at the dose with the highest desirability among the adjacent doses. In Step 2(b), the higher dose d_{j+1} is not considered due to the fact that, given that there is substantial evidence (i.e., $n_j \geq N^*$) that d_j is already close to the de-escalation boundary λ_d and d_{j+1} is likely to be overly toxic. Therefore, N^* is a cutoff for sufficiency of the data, and a larger value of N^* encourages more active exploration of new doses. By default, Lin et al. (2020b) recommended $N^* = 6$. When needed, N^* can be further calibrated by simulation to obtain certain design properties, e.g., using a larger N^* encourages faster dose exploration.

During the trial, only admissible doses can be given to incoming patients, and doses that are not admissible should be eliminated. In other words, the decision rule in Step 2 should be applied only to the doses in $\mathcal{A}(D)$. For example, in Step 2(c), if only $d_{j-1}, d_j \in \mathcal{A}(D)$ and $d_{j+1} \notin \mathcal{A}(D)$, then when we apply the rule 2(c), we should choose the level from $\{j - 1, j\}$ that has the highest RDS to treat the next cohort of patients. At any time, if all doses are eliminated, the trial should be stopped with no dose selected as OBD.

For ethical considerations, BOIN12 always assigns the next cohort of patients to the dose with the highest estimate of desirability. With a small sample size, this myopic approach may cause the dose finding to get stuck at a locally optimal dose. To alleviate that issue, a dose exploration rule can be imposed: treat the next cohort of patients at the next higher dose if the following three conditions are all satisfied:

- The number of patients treated at the current dose d_j is greater than eight.

- The observed toxicity rate $\hat{\pi}_{Tj}$ is less than the de-escalation boundary λ_d.

- The next higher dose has never been used for treating patients.

When the trial completes, the final OBD is selected based on a two-step procedure: in the first step, we select MTD based on the target toxicity rate ϕ_T so that any dose levels above the selected MTD are deemed overly toxic; in the second step, we then choose the dose level that has the highest desirability among the doses that are not higher than MTD. Specifically, at the end of the trial, we first obtain the observed marginal toxicity rates $\hat{\pi}_{Tj}$ for each dose level $j = 1, \ldots, J$. To borrow information across dose levels, an isotonic regression is performed on $\{\hat{\pi}_{Tj}\}$ through the pool-adjacent-violators algorithm

(Bril et al., 1984a), so that the isotonically transformed estimate $\tilde{\pi}_{Tj}$ mono-tonically increases with the dose. MTD, d^{MTD}, is then selected as the dose that has the estimated toxicity rate closest to the target toxicity rate ϕ_T, that is,

$$d^{\text{MTD}} = \arg\min_{d_j \in \{d_1,...,d_J\}} |\tilde{\pi}_{Tj} - \phi_T|. \tag{8.12}$$

Next, we obtain the posterior estimate of the mean utility at each dose level $j = 1, \ldots, J$ as follows:

$$\bar{u}_j = \frac{x_j + \alpha_u}{n_j + \alpha_u + \beta_u}. \tag{8.13}$$

The final OBD, d^{OBD}, is then selected as the admissible dose that does not exceed the estimated MTD, d^{MTD}, and also yields the highest estimated mean utility, that is,

$$d^{\text{OBD}} = \arg\max_{d_j \leq d^{\text{MTD}},\ d_j \in \mathcal{A}(D)} \{\bar{u}_j\}. \tag{8.14}$$

When there is a tie, select the lower dose level.

8.4.4   Operating characteristics

Lin et al. (2020b) carried out a simulation study to compare the operating characteristics of BOIN12, the model-based EffTox design, the toxicity and efficacy probability interval (TEPI) design (Li et al., 2017), and the conventional design consisting of a dose escalation phase based on the 3+3 design followed by a cohort expansion at the identified MTD (referred to as 3+3+CE). Five scenarios with the marginal toxicity and efficacy rates of each of the five dose levels considered in the simulation study are provided in Table 8.6. The highest acceptable toxicity probability is $\phi_T = 0.35$, and the lowest acceptable efficacy probability is $\phi_E = 0.25$. The utility in Table 8.1 is used to define the toxicity–efficacy trade-off. Patients are treated in cohorts of three, with a maximum sample size of 36. The Simon's two-stage design (Simon, 1989) is used to monitor efficacy in the cohort expansion for 3+3+CE.

Figure 8.5 presents the simulation results based on 10,000 simulated trials that quantify the operating characteristics of the designs, including the percentage of correct selection of OBD, the number of patients treated at OBD, the number of patients treated at the overdoses (i.e., the doses with toxicity rates > 0.35), and risk of poor allocation (i.e., the percentage of trials that allocate $< 36/5 = 7.2$ patients to OBD). In general, BOIN12 has the best overall performance among the four designs. It has the highest average percentage of correct selection of OBD and allocates the largest average number of patients to OBD. In addition, BOIN12 is safer and more reliable than other designs, as demonstrated by having amongst the smallest number of patients treated at toxic doses and the lowest risk of poor allocation.

As expected, the 3 + 3 + CE design has the worst overall performance because that design ignores the toxicity-efficacy trade-off and does not target OBD. The performance of the EffTox design varies dramatically from one scenario to another, depending on how well the assumed model reflects the true

TABLE 8.6: Five scenarios with the marginal toxicity and efficacy probabilities, as well as mean utilities (π_T, π_E, u) in the simulation study. The mean utility is based on the 2×2 utility table given in Table 8.1. The OBD that maximizes the utility is bolded.

Scenario		Dose level				
		1	2	3	4	5
1	(π_T, π_E)	$(0.05, 0.20)$	$\mathbf{(0.12, 0.35)}$	$(0.27, 0.36)$	$(0.35, 0.37)$	$(0.50, 0.38)$
	u	50.0	**56.2**	50.8	48.2	42.8
2	(π_T, π_E)	$(0.05, 0.01)$	$(0.12, 0.05)$	$(0.27, 0.30)$	$\mathbf{(0.35, 0.60)}$	$(0.50, 0.60)$
	u	38.6	38.2	47.2	**62.0**	56.0
3	(π_T, π_E)	$(0.03, 0.05)$	$(0.05, 0.10)$	$(0.20, 0.50)$	$\mathbf{(0.22, 0.68)}$	$(0.45, 0.70)$
	u	41.8	44.0	62.0	**72.0**	64.0
4	(π_T, π_E)	$(0.02, 0.05)$	$(0.05, 0.15)$	$\mathbf{(0.10, 0.40)}$	$(0.20, 0.40)$	$(0.30, 0.40)$
	u	42.2	47.0	**60.0**	56.0	52.0
5	(π_T, π_E)	$(0.15, 0.25)$	$\mathbf{(0.20, 0.55)}$	$(0.25, 0.40)$	$(0.35, 0.30)$	$(0.45, 0.20)$
	u	49.0	**65.0**	54.0	44.0	34.0

dose–toxicity or dose–efficacy relationship. Compared with BOIN12, TEPI has uniformly poor accuracy to identify OBD, a higher (i.e., more than doubled) likelihood of overdosing patients, and a higher (i.e., almost doubled) risk of poor allocation.

8.4.5 Extension to more complicated endpoints

One advantage of the quasi-binomial approach introduced in Section 8.4.1 is that it can be easily generalized to accommodate different utility (or desirability) functions. Let μ be any desirability function that can be calculated based on the data $D(d)$ observed at a specific dose d, let $\mu_{\max}$ and $\mu_{\min}$ be the upper and lower limits of μ, respectively. Then the "quasi-binomial" data $(\tilde{x}, n)$ can be defined as follows:

$$\tilde{x} = \frac{n(\mu - \mu_{\min})}{\mu_{\max} - \mu_{\min}}.$$

By substituting $(\tilde{x}, n)$ to equation (8.11), the posterior desirability distribution can be obtained and incorporated into the BOIN12 design.

As an illustration, we discuss the BOIN12 design when toxicity and efficacy endpoints have more than two levels. Assume that the efficacy outcome has three levels: complete response (CR), partial response (PR), and no response (NR); and the toxicity outcome also has three levels: no/minor, moderate, and severe. In immune checkpoint inhibitor trials, a portion of patients often experience severe immune-related adverse events (irAEs), which may need

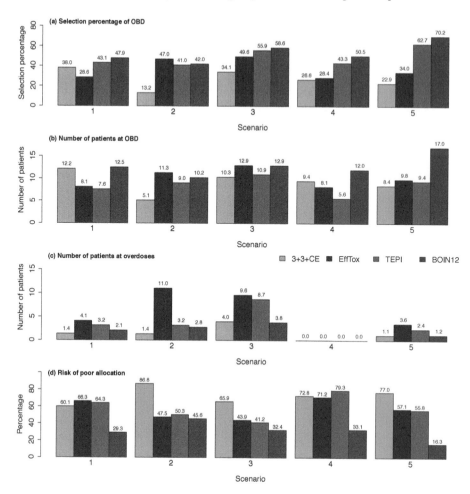

FIGURE 8.5: Percentages of correct selection of OBD, number of patients treated at OBD, number of patients treated at overdoses, and risk of poor allocations for the 3+3+CE, EffTox, TEPI, and BOIN12 designs, computed based on 10,000 simulated trials under each of the five scenarios shown in Table 8.6. The overdoses are defined as the doses with toxicity rates greater than 0.35. The maximum sample size is 36. Risk of poor allocation is the percentage of trials than allocates less than $36/5 = 7.2$ patients to OBD.

separate consideration. In this case, the "severe" level of the toxicity outcome can be used to indicate such irAEs. This results in nine possible outcomes, denoted as outcome $1, \ldots, 9$, and the corresponding utility table is given in Table 8.7.

Here, v_1 denotes the utility of the most desirable outcome (i.e., no/minor toxicity and CR), thus setting $v_1 = 100$; and v_9 denotes the utility of the most

TABLE 8.7: Utility table for three-level efficacy and toxicity endpoints.

Toxicity	Efficacy		
	CR	PR	NR
No/minor	$v_1 = 100$	v_2	v_3
Moderate	v_4	v_5	v_6
Severe	v_7	v_8	$v_9 = 0$

undesirable outcome (i.e., severe toxicity and NR), thus setting $v_9 = 0$. Using these two extreme cases as the reference, we elicit from clinicians the utility scores of the other seven possible outcomes to reflect their clinical desirability. For example, to reflect that irAE is highly undesirable, we can assign low scores to v_7 and v_8, e.g., $v_7 = 30$ and $v_8 = 10$.

Let p_l denote the probability of observing the lth outcome l at dose d, $l = 1, \ldots, 9$. The desirability (or mean utility) of dose d is given by

$$\mu = \sum_{l=1}^{9} p_l v_l,$$

and the standardized desirability is $\tilde{\mu} = (\mu - \mu_{\min})/(\mu_{\max} - \mu_{\min}) = \mu/100$. Let y_l denote the number of patients experiencing outcome l, the "quasi-binomial" outcome $\tilde{x}$ at dose d is

$$\tilde{x} = \sum_{l=1}^{9} y_l v_l / 100.$$

As a result, the quasi-binomial modelling of the utility discussed in Section 8.4.1, and also the BOIN12 dose-finding rule, can be directly applied for such a trial with three-level toxicity and three-level efficacy endpoints.

8.5 TITE-BOIN12 design

One practical issue encountered when applying BOIN12, especially in trials of targeted or immunotherapy agents, is that of possible late-onset efficacy and toxicity. As a motivating example, consider a phase I/II renal cell carcinoma trial. The objective is to determine OBD of a novel targeted agent, combined with nivolumab at a fixed dose of 3 mg/kg daily for two weeks, in patients with metastatic renal cell carcinoma. Toxicity is defined as a grade 3 or 4 liver, lung, gastrointestinal, or endocrine toxicity, or myelosuppression, within 45 days from the start of therapy according to Common Terminology Criteria for Adverse Events (CTCAE) version 5. Efficacy is defined as achieving complete response or partial response within 60 days. The expected accrual rate is three patients per month. The challenge of designing this trial is that, on average, six (or four) new patients will be enrolled by the time the previous cohort of patients complete their efficacy (or toxicity) evaluation. The question is how to treat these new patients in a timely fashion, given that the outcomes of some of the previously treated patients are still pending.

Formally, let τ_T and τ_E denote the lengths of the assessment windows for y_T and y_E, respectively. The value of τ_T and τ_E should be elicited from clinicians and wide enough to capture all necessary toxicity and efficacy events relevant to OBD determination. For example, τ_T may be the first cycle of the therapy (e.g., 21 or 28 days), while τ_E may be also the first cycle of the therapy (e.g., pharmacodynamic (PD) biomarkers), or several months after treatment (e.g., tumor response). Let t_T and t_E denote the times to toxicity and efficacy, respectively. For patients who will not experience toxicity or efficacy, define $t_T = \infty$ and $t_E = \infty$. The relation between (y_T, y_E) and (t_T, t_E) is given as follows:

$$y_T = 1\{t_T \leq \tau_T\}, \quad \text{and} \quad y_E = 1\{t_E \leq \tau_E\},$$

where $1\{\cdot\}$ denotes the indicator function, which takes a value of 1 if the event specified in the braces occurs and zero otherwise.

Suppose that n patients have been accrued and treated at dose d, we use the subscript i to indicate the data for the ith patient, $i = 1, \ldots, n$. The relationship between the multinomial variable $(y_{01}, y_{00}, y_{11}, y_{10})$ and the individual patient-level data is

$$y_{01} = \sum_{i=1}^{n} 1\left\{(y_{Ti}, y_{Ei}) = (0, 1)\right\} = \sum_{i=1}^{n} 1\left\{t_{Ti} > \tau_T, t_{Ei} \leq \tau_E\right\},$$
$$y_{00} = \sum_{i=1}^{n} 1\left\{(y_{Ti}, y_{Ei}) = (0, 0)\right\} = \sum_{i=1}^{n} 1\left\{t_{Ti} > \tau_T, t_{Ei} > \tau_E\right\},$$
$$y_{11} = \sum_{i=1}^{n} 1\left\{(y_{Ti}, y_{Ei}) = (1, 1)\right\} = \sum_{i=1}^{n} 1\left\{t_{Ti} \leq \tau_T, t_{Ei} \leq \tau_E\right\},$$
$$y_{10} = \sum_{i=1}^{n} 1\left\{(y_{Ti}, y_{Ei}) = (1, 0)\right\} = \sum_{i=1}^{n} 1\left\{t_{Ti} \leq \tau_T, t_{Ei} > \tau_E\right\}.$$

Denote t_U as the follow-up time. In cases of late-onset toxicity or efficacy, the issue is that the time to toxicity or efficacy may not be observed as fast as the patients' interarrival time. As a result, t_U may censor t_T or t_E or both of them. In other words, when t_U is smaller than t_T or t_E, y_T or y_E may be unobservable. Section 5.2 provides statistical properties of these missing data, and why we cannot simply ignore them in decision making.

As an extension of BOIN12, the TITE-BOIN12 design addresses the late-onset issue by using the follow-up times of pending patients to facilitate real-time decision making to choose the optimal dose for new patients (Zhou et al., 2022). More precisely, by treating the outcome data of the pending patients as missing, two approaches, based on different imputation techniques, have been proposed: Bayesian data augmentation (BDA) (Little and Rubin, 2014) and mean imputation based on the approximate likelihood (Lin and Yuan, 2020).

Bayesian data augmentation
BDA iterates between two steps: the imputation (I) step, in which the missing y_T and y_E are imputed from their conditional posteriors, and the posterior

(P) step, in which the posterior samples of unknown parameters including $(p_{01}, p_{00}, p_{11}, p_{10})$ and $\tilde{u}$ are simulated based on the imputed data. As the P step is the same as BOIN12, we here focus on the I step, which involves the conditional posteriors of missing y_T and y_E.

In the presence of late-onset toxicity or efficacy, there are three possible missing patterns: (1) both y_T and y_E are missing, (2) only y_T is missing, and (3) only y_E is missing. Define $\delta_T = 1\{t_U < \tau_T \cap t_U < t_T\}$ as the missing indicator for toxicity, which takes a value of 1 if y_T is missing. Similarly, define $\delta_E = 1\{t_U < \tau_E \cap t_U < t_E\}$ as the missing indicator for efficacy. The imputation steps for the three missing patterns are described as follows:

- When both y_T and y_E are missing, we impute (y_T, y_E) by drawing a random sample from the multinomial distribution with individual joint probabilities given by

$$\Pr\{(y_T, y_E) = (a, b) \mid (\delta_T, \delta_E) = (1, 1)\} = \frac{p_{ab} S_{ab}}{\sum_{a'=0}^{1} \sum_{b'=0}^{1} p_{a'b'} S_{a'b'}},$$

for $a, b \in \{0, 1\}$, where $S_{ab} = \Pr\{t_T > t_U, t_E > t_U \mid (y_T, y_E) = (a, b)\}$. Assuming "working" independence between t_T and t_E, we have

$$S_{ab} = \Pr(t_T > t_U \mid y_T = a) \Pr(t_E > t_U \mid y_E = b).$$

Let $w_q = \Pr(t_q \leq t_U \mid y_q = 1)$ denote a weight, which accounts for the partial information from patients who have not completed the assessment of y_q, $q \in \{T, E\}$. The specification of w_q will be discussed below. As $\Pr(t_q > t_U \mid y_q = 0) = 1$, $q \in \{T, E\}$, we have $S_{00} = 1$, $S_{01} = 1 - w_E$, $S_{10} = 1 - w_T$, and $S_{11} = (1 - w_T)(1 - w_E)$.

- When y_T is observed but y_E is missing, we draw the missing value of y_E from a Bernoulli distribution with the response probability given by

$$\Pr(y_E = 1 \mid y_T, \delta_E = 1) = \left\{ \frac{p_{11}(1 - w_E)}{p_{10} + p_{11}(1 - w_E)} \right\}^{y_T} \left\{ \frac{p_{01}(1 - w_E)}{p_{00} + p_{01}(1 - w_E)} \right\}^{1 - y_T}.$$

- When y_E is observed but y_T is missing, we draw missing y_T from a Bernoulli distribution with the toxicity probability given by

$$\Pr(y_T = 1 \mid y_E, \delta_T = 1) = \left\{ \frac{p_{11}(1 - w_T)}{p_{01} + p_{11}(1 - w_T)} \right\}^{y_E} \left\{ \frac{p_{10}(1 - w_T)}{p_{00} + p_{10}(1 - w_T)} \right\}^{1 - y_E}.$$

Derivation for the above conditional probabilities and more details of the BDA procedure are provided in Zhou et al. (2022). Following Cheung and Chappell (2000) and Yuan et al. (2018), the TITE-BOIN12 design by default assumes that the time to toxicity and time to efficacy are uniformly distributed over the respective assessment windows. As a result, $w_T = t_U/\tau_T$, and $w_E = t_U/\tau_E$. This assumption seems strong, but it is very robust for the purpose of dose finding (Cheung and Chappell, 2000; Yuan et al., 2018; Zhou et al., 2022).

Mean imputation based on the approximated likelihood

One limitation of BDA is that it is computationally intensive. A second approach is to use mean imputation to deal with the missing data through the use of the approximated likelihood approach by Lin and Yuan (2020). The key observation is that, in the presence of pending outcomes, the reason that BOIN12 cannot be directly applied is that the quasi-number of events x given by equation (8.10) cannot be calculated. Zhou et al. (2022) proposed to approximate x, and thus the quasi-binomial likelihood, by replacing the pending/missing values of y_T and y_E with their expectations $E(y_T \mid \delta_T = 1) = \Pr(y_T = 1 \mid \delta_T = 1)$ and $E(y_E \mid \delta_E = 1) = \Pr(y_E = 1 \mid \delta_E = 1)$, respectively.

Depending on the values of δ_T and δ_E, patients can be divided into four types, i.e., $(\delta_T, \delta_E) = (0, 0), (1, 0), (0, 1), (1, 1)$. The quasi-number of events x can be approximated as follows,

$$
\begin{aligned}
x \;=\; & \frac{1}{100} \sum_{a=0}^{1} \sum_{b=0}^{1} \Bigg\{ v_{ab} \sum_{i=1}^{n} 1\{y_{Ti} = a\} 1\{y_{Ei} = b\} (1 - \delta_{Ti})(1 - \delta_{Ei}) \\
& + v_{ab} \sum_{i=1}^{n} 1\{y_{Ti} = a\} \Pr(y_{Ei} = b \mid \delta_{Ei} = 1)(1 - \delta_{Ti})\delta_{Ei} \\
& + v_{ab} \sum_{i=1}^{n} \Pr(y_{Ti} = a \mid \delta_{Ti} = 1) 1\{y_{Ei} = b\} \delta_{Ti}(1 - \delta_{Ei}) \\
& + v_{ab} \sum_{i=1}^{n} \Pr(y_{Ti} = a \mid \delta_{Ti} = 1) \Pr(y_{Ei} = b \mid \delta_{Ei} = 1)\delta_{Ti}\delta_{Ei} \Bigg\}.
\end{aligned}
$$

In this equation, the first term corresponds to those patients whose y_T and y_E are both observed (i.e., $(\delta_T, \delta_E) = (0, 0)$), thus their contribution to the quasi-number of events x can be directly computed based on the observed data. The last three terms correspond to the patients with at least one of y_T and y_E pending, i.e., $\delta_T + \delta_E > 0$, and involve the mean imputation of y_T and y_E respectively by $\Pr(y_T = a \mid \delta_T = 1)$ and $\Pr(y_E = 1 \mid \delta_E = b)$, $a, b = 0, 1$, based on the assumption that y_T and y_E are "working" independent. Technically speaking, the correlation between efficacy and toxicity can be modeled. The small sample sizes of phase I/II trials, however, provide limited information to estimate the correlation parameter reliably. The independence assumption for pending outcomes seems strong, but it makes the method simple and computationally fast. Zhou et al. (2022) showed by numerical studies that the TITE-BOIN12 design is remarkably robust to the violation of this assumption. This may be because the assumption is only used for the patients who have pending data. For patients with both y_T and y_E observed, the independence assumption is not made. When the trial progresses, the percentage of observed data increases, limiting the impact of the violation of the independence assumption.

By assuming that given $y_q = 1$, the time-to-event outcome t_q is a uniform random variable over $(0, \tau_q)$, we have $\Pr(\delta_q = 1 \mid y_q = 1) = \Pr(t_q > t_U \mid y_q =$

1) $= 1 - t_U/\tau_q$, and thus

$$\Pr(y_q = 1 \mid \delta_q = 1)$$
$$= \frac{\Pr(\delta_q = 1 \mid y_q = 1)\Pr(y_q = 1)}{\Pr(\delta_q = 0 \mid y_q = 0)\Pr(y_q = 0) + \Pr(\delta_q = 1 \mid y_q = 1)\Pr(y_q = 1)}$$
$$= \frac{\Pr(\delta_q = 1 \mid y_q = 1)\Pr(y_q = 1)}{\Pr(y_q = 0) + \Pr(\delta_q = 1 \mid y_q = 1)\Pr(y_q = 1)}$$
$$= \frac{\pi_q(1 - t_U/t_q)}{1 - \pi_q t_U/t_q},$$

for $q = T, E$. As a result, $\Pr(y_q = 0 \mid \delta_q = 1) = 1 - \Pr(y_q = 1 \mid \delta_q = 1) = (1 - \pi_q)/(1 - \pi_q t_U/t_q)$.

To further calculate the unknown π_q involved in the above equation, the approximated likelihood approach by Lin and Yuan (2020) is adopted. As derived in Section 5.5, the marginal likelihood function for a patient with outcome q observed is $\pi_q^{y_q}(1 - \pi_q)^{1-y_q}$, and that for a patient with outcome q pending is $1 - \pi_q w_q$, where $w_q = \Pr(t_q \le t_U \mid y_q = 1) = t_U/\tau_q$. Based on the approximation $1 - \pi_q w_q \approx (1 - \pi_q)^{w_q}$, the (marginal) overall likelihood function for π_q based on the data $D(d)$ observed at a specific dose d is given by

$$
\begin{aligned}
L(D(d) \mid \pi_q) &= \prod_{i=1}^{n} \left\{ \pi_q^{y_{qi}}(1 - \pi_q)^{1-y_{qi}} \right\}^{1-\delta_{qi}} (1 - \pi_q w_{qi})^{\delta_{qi}} \\
&\approx \prod_{i=1}^{n} \left\{ \pi_q^{y_{qi}}(1 - \pi_q)^{1-y_{qi}} \right\} (1 - \pi_q)^{\delta_{qi} w_{qi}} \\
&= \pi_q^{\tilde{n}_q}(1 - \pi_q)^{\tilde{m}_q},
\end{aligned}
$$

where $\tilde{n}_q = \sum_{i=1}^{n}(1-\delta_{qi})y_{qi}$ is the number of outcomes for endpoint q observed so far by the interim time, and $\tilde{m}_q = \sum_{i=1}^{n}(1 - \delta_{qi})(1 - y_{qi}) + \sum_{i=1}^{n} \delta_{qi} w_{qi}$ is the "effective" number of patients who do have endpoint q, $q = T, E$.

Various appropriate estimates of the unknown π_q can be easily obtained based on the above approximated likelihood function. For example, the most straightforward one is the maximum likelihood estimator $\hat{\pi}_q = \tilde{n}_q/(\tilde{n}_q + \tilde{m}_q), q = T, E$. By plugging $\hat{\pi}_q$ into the expressions of $\Pr(y_q = 1 \mid \delta_q = 1)$ and $\Pr(y_q = 0 \mid \delta_q = 1)$, the quasi-number of events x can be calculated accordingly. Then, the method of BOIN12 can be directly applied to obtain the posterior of $\tilde{u}$ as in equation (8.11) for decision making.

The dose-finding algorithm for TITE-BOIN12, using either BDA or approximated likelihood methods, is the same as that for BOIN12. At each interim decision, TITE-BOIN12 needs to update the admissible set $\mathcal{A}(D)$, the estimate for the marginal toxicity probability $\hat{\pi}_T$ at the current dose, and the desirability PP_j for each dose considered in the trial. In addition, to avoid risky decisions caused by sparse data, TITE-BOIN12 imposes the following accrual suspension rule: if more than 50% of the patients have pending DLT or efficacy outcomes at the current dose, suspend the accrual to wait for more data to become available.

Of note, another appealing feature of TITE-BOIN12 is that it naturally accommodates the case that some patients may be evaluable for toxicity, but not evaluable for efficacy (e.g., patients are off treatment due to toxicity). These patients can be regarded as y_T are observed and y_E are (permanently) pending, and thus directly incorporated into the utility estimation and decision making.

8.6   Other model-assisted phase I/II designs

8.6.1   uTPI design

Similar to the BOIN12 design, the uTPI design (Shi et al., 2021) also adopts the Dirichlet-multinomial and quasi-binomial approaches to model the joint efficacy–toxicity and quasi-binomial utility outcomes, respectively. Unlike BOIN12, whose decisions are made based on the point estimate of the toxicity rate and the posterior probability distribution of the utility, the uTPI design is a seamless extension of the keyboard design (a.k.a., the m-TPI2 design), and it uses the toxicity and utility intervals in determining the optimal dose for the incoming patients.

Specifically, following the idea of the keyboard design, the uTPI design starts by partitioning the support of the probability of toxicity $(0, 1)$ into a series of intervals of equal width ϵ, which results in K_T toxicity intervals, $\mathcal{I}_{T,k}, k = 1, \ldots, K_T$. Similarly, the utility support $(0, 100)$ also can be partitioned into K_U intervals, $\mathcal{I}_{U,k}, k = 1, \ldots, K_U$, based on an equal width δ. The parameter ϵ (or δ) denotes the indifference margin for toxicity (or utility), which means that any two dose levels whose toxicity probabilities (or mean utilities) lying within the same toxicity (or utility) interval are treated indifferently in terms of toxicity or mean utility. The reason for specifying the indifference intervals is that the sample size in early-phase dose-finding trials is typically small, and thus it is difficult to differentiate two dose levels that have similar toxicity probabilities or mean utilities.

For illustration, we take $\epsilon = 0.1$ and $\delta = 10$, leading to 10 toxicity intervals $\mathcal{I}_{T,1} = (0, 0.1), \ldots, \mathcal{I}_{T,10} = (0.9, 1)$ and 10 utility intervals $\mathcal{I}_{U,1} = (0, 10), \ldots, \mathcal{I}_{T,10} = (90, 100)$. By combining the toxicity and utility intervals, the equal-width "keys" in a one-dimensional keyboard design are extended to equal-area "squares" on a two-dimensional chessboard, as shown in Figure 8.6.

Mimicking the keyboard design, the strongest toxicity interval that possesses the largest posterior probability can be identified based on data $D(d)$

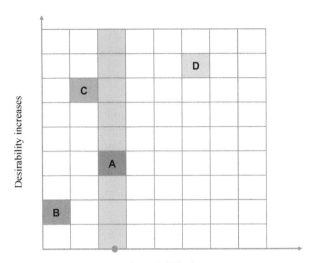

Toxicity probability increases

FIGURE 8.6: Dose-finding rule of the uTPI design based on a chessboard. The shaded interval corresponds to the target toxicity probability interval k^*. (1) If C is the current dose (or if A is the current dose and $n(A) < N^* = 9$), and suppose that B is the next lower dose and D is the next higher dose, we will allocate the next cohort of patients to the dose among $\{C, B, D\}$ (or $\{A, B, D\}$) that has the highest utility interval. In this case, we will select dose D for the next cohort of patients. (2) If A is the current dose and $n(A) \geq N^* = 9$, and suppose that B is the next lower dose and D is the next higher dose, we will allocate the next cohort of patients to the dose among $\{A, B\}$ that has the highest utility interval. In this example, dose A will be selected for treating the next cohort of patients.

observed at dose d,

$$k_d^T = \underset{k=1,\dots,K_T}{\arg\max} \left\{ \Pr\left(\pi_T(d) \in \mathcal{I}_{T,k} \mid D(d) \right) \right\},$$

where the posterior probability distribution of $\pi_T(d)$ is derived under the beta-binomial model (8.6). Similarly, the strongest utility interval that has the largest posterior probability can be obtained by the quasi-beta-binomial model (8.11),

$$k_d^U = \underset{k=1,\dots,K_U}{\arg\max} \left\{ \Pr\left(\pi_U(d) \in \mathcal{I}_{U,k} \mid D(d) \right) \right\}.$$

The collection of strongest toxicity and desirability intervals (k_d^T, k_d^U) can be treated as the vector of sufficient statistics of the uTPI design in determining dose escalation or de-escalation. Specifically, let k^* denote the location of the toxicity interval that contains the upper limit of the toxicity rate ϕ_T. In the aforementioned example, $k^* = 4$ when $\phi_T = 0.35$. As demonstrated in Figure 8.6, the dose-finding rule of uTPI proceeds as follows:

1. Patients in the first cohort are treated at the lowest dose d_1, or the physician-specified dose.

2. Based on the observed data, we obtain the strongest toxicity and desirability intervals for each dose. Suppose j is the current dose level. To assign a dose to the next cohort of patients,

    (a) If $k_{d_j}^T > k^*$, de-escalate the dose to level $j - 1$.

    (b) If $k_{d_j}^T = k^*$ and $n_j \geq N^*$, choose the level from $\{j - 1, j\}$ that has a larger utility interval k_d^U.

    (c) Otherwise, if $k_{d_j}^T < k^*$ or $k_{d_j}^T = k^*$ and $n_j < N^*$, choose the level from $\{j - 1, j, j + 1\}$ that has the largest utility interval k_d^U.

3. Repeat Step 2 until the maximum sample size N is reached, and then use the OBD selection rule described in Section 8.4.3 to select the OBD.

During the trial conduct, uTPI only treats the patients using doses in the admissible set $\mathcal{A}(D)$ as defined in BOIN12 and U-BOIN. If no dose is admissible, the trial should be terminated early.

Similar to BOIN12 and U-BOIN designs, the uTPI design also has a concise decision structure such that a dose-assignment decision table can be calculated before the trial starts and can be used throughout the trial, which simplifies its practical implementation. More details about the implementation of the uTPI design can be found in Shi et al. (2021).

8.6.2   STEIN and BOIN-ET designs

In some trials, OBD is defined as the lowest safe dose that has the maximum efficacy, instead of the safe dose that has the highest utility. In general, the designs described previously can be readily adapted to these trials simply by properly specifying the utility functions, e.g., setting $v_{11} = 100$ and $v_{00} = 0$ in the BOIN12, U-BOIN, and uTPI designs. Without using utility, some model-assisted designs have been proposed to directly find the OBD that has the maximum efficacy.

Lin and Yin (2017b) proposed the STEIN design on the basis of the standard BOIN design. Besides using BOIN's toxicity-based rule to determine whether a dose is safe or not, the STEIN design additionally uses the hypotheses stated next to determine whether the efficacy at a dose is acceptable or not. Specifically, at each dose d_j, two point hypotheses are defined:

$$H_{0j}^E : \pi_E(d_j) = \psi_1 \quad \text{versus} \quad H_{1j}^E : \pi_E(d_j) = \psi_2,$$

where ψ_1 denotes a clinically uninteresting response rate, and ψ_2 denotes a clinically desired response rate. Let $\hat{\pi}_E(d_j) = n_E(d_j)/n(d_j)$ be the observed efficacy rate at dose d_j. The null hypothesis H_{0j}^E is rejected and dose d_j is considered to be efficacious if $\hat{\pi}_E(d_j) \geq \delta$, where δ is the boundary for the observed efficacy rate. For ease of exposition, we slightly abuse the notation and use expanded notations such as $\hat{\pi}_E(d_j)$, $n_E(d_j)$ and $n(d_j)$ to represent the same things as $\hat{\pi}_{Ej}$, n_{Ej} and n_j used previously.

Based on a similar procedure of BOIN, the optimal boundary of the observed efficacy rate can be obtained,

$$\delta = \frac{\log\left(\dfrac{1-\psi_1}{1-\psi_2}\right)}{\log\left(\dfrac{\psi_2(1-\psi_1)}{\psi_1(1-\psi_2)}\right)},$$

where non-informative prior is assumed with $\Pr(H_{0j}^E) = \Pr(H_{1j}^E) = 1/2$.

By partitioning the two-dimensional efficacy–toxicity space into several regions (Figure 8.7), the dose-finding rule of STEIN can be described as follows:

1. Patients in the first cohort are treated at the lowest dose d_1, or the physician-specified dose.

2. Suppose j is the current dose level. To assign a dose to the next cohort of patients:

   (a) If $\hat{\pi}_T(d_j) > \lambda_d$, de-escalate the dose to level $j - 1$.

   (b) If $\hat{\pi}_T(d_j) \leq \lambda_d$ and $\hat{\pi}_E(d_j) \geq \delta$, stay at the current dose.

   (c) Otherwise, choose the dose that has the largest posterior probability of $\Pr\left(\pi_E(d) > \delta \mid D(d)\right)$, which can be obtained based on the beta-binomial model (8.6).

3. Repeat Step 2 until the maximum sample size N is reached.

To use the STEIN design in practice, Lin and Yin (2017b) recommended setting $\phi_1 = 0.75\phi$ and $\phi_2 = 1.25\phi$ to obtain BOIN's escalation/de-escalation boundaries, and they found that $\psi_1 \in (0.2, 0.4)$ and $\psi_2 \in (0.6, 0.8)$ are suitable for most practical scenarios. For example, when $\phi = 0.3$, $\psi_1 = 0.3$, and $\psi_2 = 0.8$, the values of $(\lambda_e, \lambda_d, \delta)$ are $(0.26, 0.34, 0.56)$. When appropriate, these design parameters can be further calibrated by simulation.

Similar to the STEIN design, the BOIN-ET design is another extension of the BOIN design to identify OBD based on both efficacy and toxicity outcomes (Takeda et al., 2018). Suppose the current dose level is j, according to Figure 8.8, the next dose assignment of BOIN-ET is as follows:

(1) If $\hat{\pi}_T(d_j) \leq \lambda_e$ and $\hat{\pi}_E(d_j) \leq \delta$, then the current dose is safe but not effective, and the next dose should be escalated to level $j + 1$.

(2) If $\hat{\pi}_T(d_j) > \lambda_d$, then the current dose is too toxic and the next dose should be de-escalated to level $j - 1$.

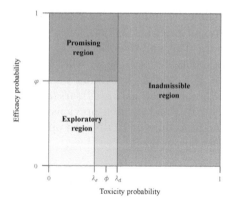

FIGURE 8.7: Partitioned regions under the STEIN design. If the pair of the observed toxicity and efficacy probabilities at the current dose level lies inside the promising region, the current dose is retained. If the pair lies inside the exploratory region, then more dose levels should be explored. Specifically, the lighter color subregion indicates that the current dose is safe and thus dose escalation is warranted. The dark color subregion indicates that the toxicity probability is close to the target, and thus the decision of dose escalation/de-escalation depends on the observed toxicity and efficacy data jointly. If the pair lies inside the inadmissible region, the current dose is too toxic and thus dose de-escalation is needed.

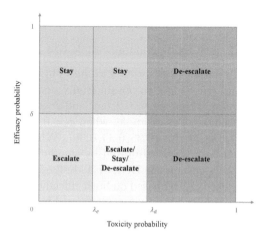

FIGURE 8.8: Dose allocation rules for the BOIN-ET design considering both efficacy and toxicity.

(3) If $\hat{\pi}_T(d_j) < \lambda_d$ and $\hat{\pi}_E(d_j) > \delta$, then the current dose is desirable in terms of both efficacy and safety, then the next dose should not be changed.

(4) If $\lambda_e < \hat{\pi}_T(d_j) \leq \lambda_d$ and $\hat{\pi}_E(d_j) \leq \delta$, then all possible decisions including escalation, stay, and de-escalation are considered. Typically, set the admissible set $\mathcal{A}_{j.} = \{j-1, j, j+1\}$, the next dose is determined according to the following rules:

(a) If dose level $j+1$ is untried, then the next dose is escalated to level $j+1$.

(b) Otherwise, the next dose is selected as the one with the maximum observed efficacy rate. If there is a tie having the maximum observed efficacy rate, then the next dose is randomly chosen from the tied doses.

To determine the optimal values of $(\lambda_e, \lambda_d, \delta)$, Takeda et al. (2018) considered the following six hypotheses at dose d_j:

$$H_{1j} : \pi_T(d_j) = \phi_1, \pi_E(d_j) = \psi_1, \quad H_{2j} : \pi_T(d_j) = \phi_1, \pi_E(d_j) = \psi_2,$$
$$H_{3j} : \pi_T(d_j) = \phi, \pi_E(d_j) = \psi_1, \quad H_{4j} : \pi_T(d_j) = \phi, \pi_E(d_j) = \psi_2,$$
$$H_{5j} : \pi_T(d_j) = \phi_2, \pi_E(d_j) = \psi_1, \quad H_{6j} : \pi_T(d_j) = \phi_2, \pi_E(d_j) = \psi_2,$$

where the design parameters $(\phi, \phi_1, \phi_2, \psi_1, \psi_2)$ have similar interpretations as in the STEIN design. By treating efficacy and toxicity jointly, the BOIN-ET design finds the optimal values that minimize the probability of incorrect decisions using a numerical grid search. More details can be found in Takeda et al. (2018).

To implement the BOIN-ET design, Takeda et al. (2018) recommended setting $\phi_1 = 0.1\phi$, $\phi_2 = 1.4\phi$, $\psi_1 = 0.6\psi_2$, and assuming $\pi_{1j} = \cdots = \pi_{6j} = 1/6$. For illustration, consider the target toxicity and efficacy rates are $(\phi, \psi_2) = (0.3, 0.6)$, the optimal values of $(\lambda_e, \lambda_d, \delta)$ are calculated to be $(0.14, 0.35, 0.48)$. To accommodate potentially late-onset toxicity or efficacy, Takeda et al. (2020) extended BOIN-ET to TITE-BOIN-ET along a similar line as TITE-BOIN12 using the approximated likelihood approach.

8.7   Software and case study

Software

The BOIN12, TITE-BOIN12, and U-BOIN designs can be implemented using web applications provided at http://www.trialdesign.org. The software allows users to generate the decision table, evaluate operating characteristics of the design, and generate the trial design template for protocol preparation. The software for the EffTox design is freely available at the MD Anderson Cancer Center Software Download Website https://biostatistics.mdanderson.org/softwaredownload/. The R code for simulating the uTPI design is available at `https://github.com/ruitaolin/uTPI`.

Case study

Platinum-Refractory Oral Cancer Trial   The objective of this phase I–II trial (Clinical Trials Registry-India identifier: CTRI/2019/01/016837) is to determine OBD for methotrexate when given along with erlotinib and celecoxib in treating patients with platinum-resistant or early-failure squamous cell carcinoma of the oral cavity. DLT is scored based on the Common Terminology Criteria for Adverse Events (version 4.03), and the efficacy endpoint used in monitoring is the clinical benefit rate at two months. Five doses of methotrexate (i.e., 3, 6, 9, 12, and 15 mg/m^2) are investigated in combination with 150 mg erlotinib and 200 mg celecoxib.

Additional examples include a CD33 CAR T-cell therapy trial in patients with relapsed or refractory acute myeloid leukemia (ClinicalTrials.gov Identifier: NCT04835519, based on BOIN12), and a trial that evaluates a plant extract to treat breast cancer patients (ClinicalTrials.gov Identifier: NCT05007444, based on U-BOIN).

We use the oral cancer trial as an example to illustrate the use of the BOIN12 app to design phase I-II trials. The BOIN12 app can be selected and launched from the BOIN Suite launchpad (Figure 8.9). The app includes six

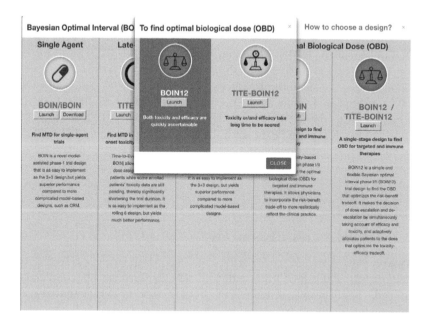

FIGURE 8.9: The launchpad of web application "BOIN Suite" available at `http://www.trialdesign.org`.

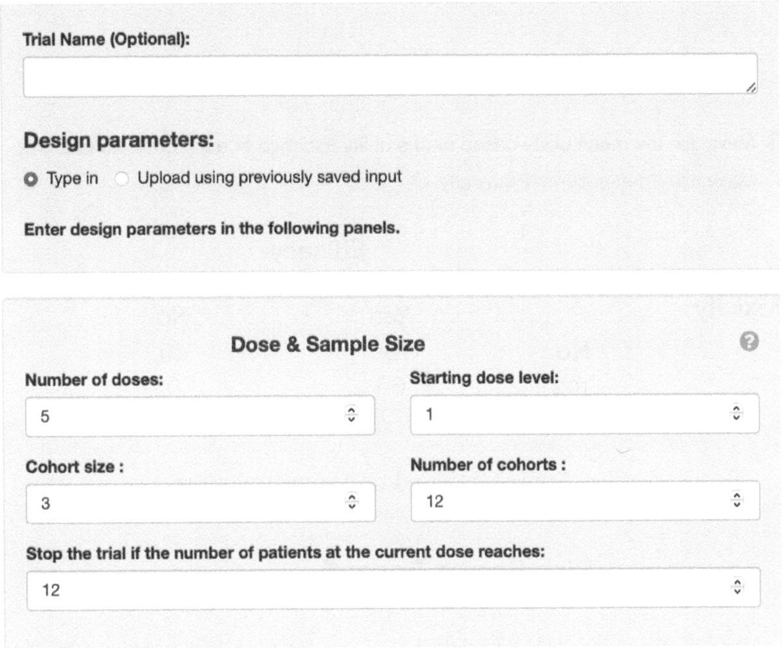

FIGURE 8.10: Specify doses and the sample size.

functional tabs and has the capability to implement the three main tasks: trial design, trial conduct, and OBD determination. We design the trial using the following three steps:

Step 1: Enter trial parameters

Doses and Sample Size   As shown in Figure 8.10, the number of doses under investigation for methotrexate is five, and the starting dose level is 1. The total sample size is 36 and patients are treated in the cohort size of three, which results in a total of 12 cohorts. We recommend the maximum sample size N should be no less than $6 \times J$ (i.e., the maximum sample size of the 3+3 design), where J is the number of doses. To reduce the sample size, the "convergence" stopping rule is implemented in BOIN12, that is, when m patients have been treated by the current dose and the decision is to stay, the trial should be early stopped before the exhaustion of the maximum sample size. This stopping criterion indicates that the trial can be stopped when the dose finding approximately converges to OBD. In this trial, $m = 12$ is used. Because of this early stopping rule, the actual sample size used in the trial is often smaller than N. The saving in sample size depends on the underlying dose–toxicity or dose–efficacy scenario and can be evaluated through simulation studies.

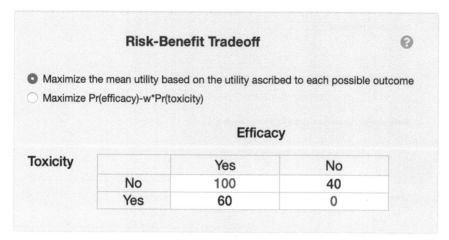

(a) Approach based on the utility score.

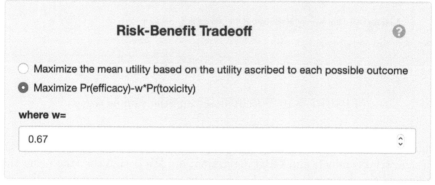

(b) Approach based on toxicity and efficacy probabilities.

FIGURE 8.11: Two approaches to specify the risk–benefit trade-off based on (a) utility scores for possible outcomes or (b) marginal toxicity and efficacy probabilities.

Risk-benefit Trade-off Criteria   In Figure 8.11 (a), the utility is used to define the risk-benefit trade-off (or desirability), where the outcome (toxicity, efficacy) has a slightly higher utility score than (no toxicity, no efficacy). Utility should be elicited from physicians to reflect the risk-benefit trade-off underlying their clinical decision making. After specifying the utility, simulation should be performed to evaluate the operating characteristics of the design. In some cases, the simulation results may motivate slight modification of some of the numerical utility values, although such modification typically has little or no effect on the design's operating characteristics. Given the utility table, the desirability of a dose is characterized by the mean utility (8.9), and a higher value of the mean utility means the dose is more desirable.

The software provides another option to specify the risk-benefit trade-off based on the marginal toxicity and efficacy probabilities, see Figure 8.11 (b). As described in Section 8.3.1, this approach actually is a special case of the utility approach with $v_{00} + v_{11} = 100$. For example, the utility specified in Figure 8.11 (a) is equivalent to specifying the toxicity-efficacy trade-off as Pr(efficacy)-2/3Pr(toxicity). In other words, the risk-benefit trade-off specified in Figure 8.11 (a) and (b) are equivalent.

In some trials, the objective is to find the dose that is safe and has the highest efficacy rate. This can be implemented by (1) setting $v_{11} = 100$ and $v_{00} = 0$ in the utility table, or (2) set $w = 0$ in the toxicity–efficacy probability trade-off function. In both cases, the safety condition is imposed through the Admissible Criteria, as discussed next.

Admissible Criteria   To safeguard patients from toxic and/or futile doses, two dose acceptability criteria discussed in Section 8.4.2 are used by BOIN12 to determine which doses may be used to treat patients. According to Figure 8.12, the highest acceptable toxicity probability is $\phi_T = 0.35$, and the lowest acceptable efficacy probability is $\phi_E = 0.25$. Generally, ϕ_E can take the value of the target response rate specified for a standard phase II trial. Because U-BOIN considers the toxicity–efficacy trade-off, the value of ϕ_T should be set slightly higher (e.g., 0.05) than the target toxicity rate used in conventional toxicity-based phase I designs. For example, if 30% is an appropriate target toxicity rate that the conventional phase I design used, then $\phi_T = 0.35$ is a reasonable choice for U-BOIN.

For the toxicity and efficacy probability cutoffs, we recommend $c_T = 0.95$ and $c_E = 0.90$, which seem high, but actually are appropriate as their purpose

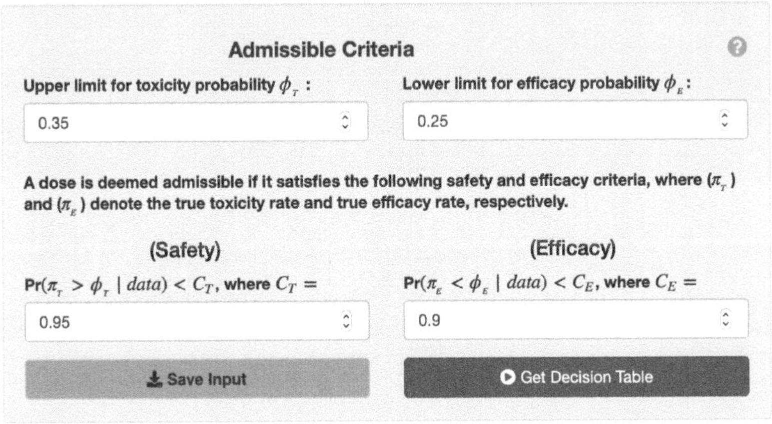

Admissible Criteria   ❓

Upper limit for toxicity probability ϕ_T :   Lower limit for efficacy probability ϕ_E:

| 0.35 ⬍ | | 0.25 ⬍ |

A dose is deemed admissible if it satisfies the following safety and efficacy criteria, where (π_T) and (π_E) denote the true toxicity rate and true efficacy rate, respectively.

(Safety)   **(Efficacy)**

$\Pr(\pi_T > \phi_T \mid data) < C_T$, where $C_T =$ $\Pr(\pi_E < \phi_E \mid data) < C_E$, where $C_E =$

| 0.95 ⬍ | | 0.9 ⬍ |

⬇ Save Input   ▶ Get Decision Table

FIGURE 8.12: Admissible criteria of BOIN12 design.

is to rule out excessively toxic and ineffective doses. Among admissible doses, the dose assignment rule will allocate patients to the most desirable dose. In other words, even if the admissible dose set includes some doses that are not particularly safe or efficacious, the design will not assign patients to these suboptimal doses. Due to the large uncertainty of small sample sizes, using small values for c_T and c_E will inadvertently eliminate the doses that are actually admissible, and thus affect the operating characteristics of the design. If a dose is inadmissible due to violation of the safety criterion, this dose and its higher doses are considered inadmissible. During the trial, only admissible doses can be used to treat patients, and the doses that are not admissible should be eliminated from the trial.

After the completion of the specification of design parameters, the design flowchart (Figure 8.13) and decision tables (Figure 8.14) can be generated by clicking the "Get Decision Table" button (see Figure 8.12). In BOIN12, two decision tables will be generated: the first one summarizes the

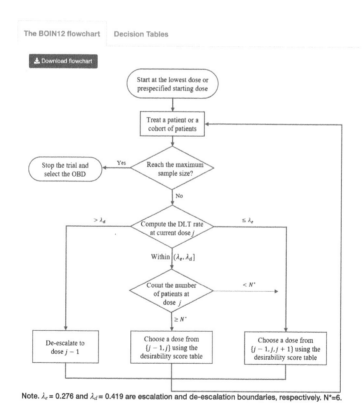

Note. λ_e = 0.276 and λ_d = 0.419 are escalation and de-escalation boundaries, respectively. N*=6.

FIGURE 8.13: The flowchart of the BOIN12 design generated by the BOIN12 shiny app.

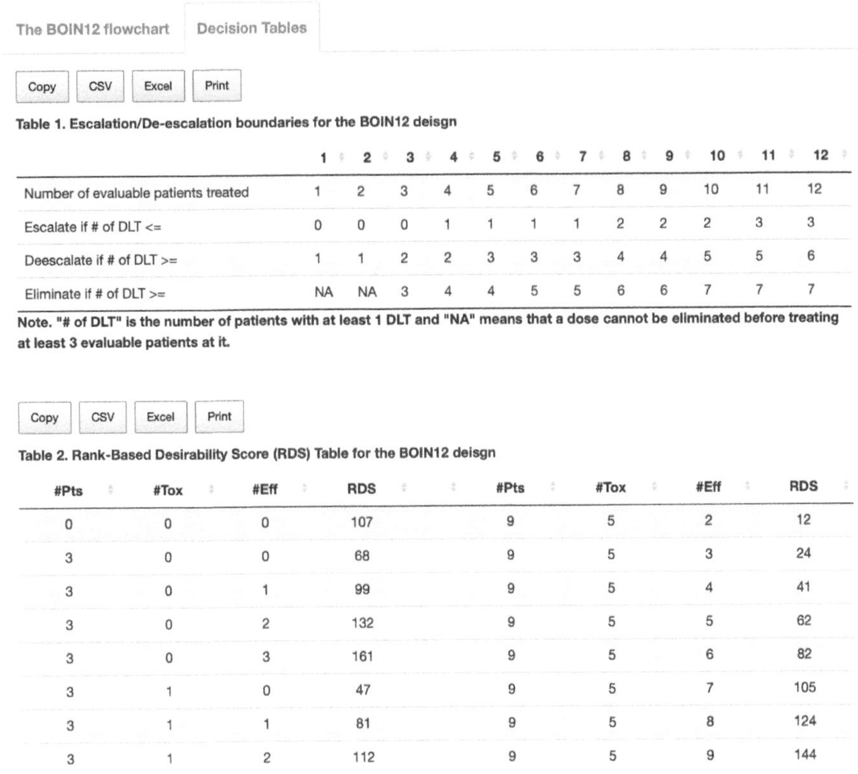

The BOIN12 flowchart | Decision Tables

| Copy | CSV | Excel | Print |

Table 1. Escalation/De-escalation boundaries for the BOIN12 deisgn

	1	2	3	4	5	6	7	8	9	10	11	12
Number of evaluable patients treated	1	2	3	4	5	6	7	8	9	10	11	12
Escalate if # of DLT <=	0	0	0	1	1	1	1	2	2	2	3	3
Deescalate if # of DLT >=	1	1	2	2	3	3	3	4	4	5	5	6
Eliminate if # of DLT >=	NA	NA	3	4	4	5	5	6	6	7	7	7

Note. "# of DLT" is the number of patients with at least 1 DLT and "NA" means that a dose cannot be eliminated before treating at least 3 evaluable patients at it.

| Copy | CSV | Excel | Print |

Table 2. Rank-Based Desirability Score (RDS) Table for the BOIN12 deisgn

#Pts	#Tox	#Eff	RDS		#Pts	#Tox	#Eff	RDS
0	0	0	107		9	5	2	12
3	0	0	68		9	5	3	24
3	0	1	99		9	5	4	41
3	0	2	132		9	5	5	62
3	0	3	161		9	5	6	82
3	1	0	47		9	5	7	105
3	1	1	81		9	5	8	124
3	1	2	112		9	5	9	144

FIGURE 8.14: The escalation/de-escalation boundary table (upper panel) and rank-based desirability score table (lower panel) of the BOIN12 design generated by the BOIN12 shiny app.

escalation/de-escalation boundaries for toxicity monitoring, and the second one is the RDS table that defines the desirability of each dose. The decision tables will be automatically included in the protocol template in Step 3, but can also be saved as separate csv, Excel, or pdf files in this step when needed.

Step 2: Run simulation

Operating Characteristics   This step generates the operating characteristics of the design through simulation, see Figure 8.15. Scenarios used for simulation should cover various possible clinical scenarios, e.g., OBD located at different dose levels. The software uses the latent bivariate normal random variables to generate the joint toxicity and efficacy outcomes. Users need to provide the true marginal toxicity rate and efficacy rate, as well as the correlation between toxicity and efficacy to simulate data. Here, the correlation corresponds to the correlation parameter in the covariance matrix of the latent bivariate normal

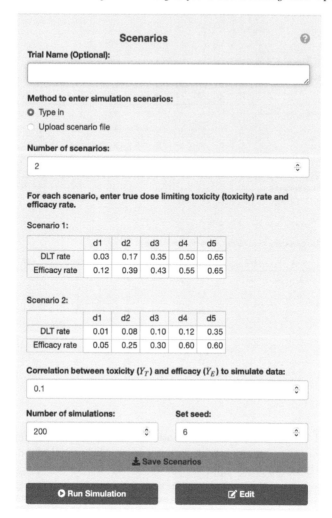

FIGURE 8.15: Simulate the operating characteristics the BOIN12 design.

distribution. The simulation results (shown in Table 8.8) will be automatically included as a table in the protocol template in the next step, but can also be saved as a separate csv or Excel file if needed.

Step 3: Generate protocol template

Protocol Preparation   The BOIN12 software generates sample texts and a protocol template to facilitate the protocol write-up. The protocol template can be downloaded in various formats (see Figure 8.16). Use of this module requires the completion of Steps 1 and 2. Once the protocol is approved by regulatory bodies (e.g., Institutional Review Board), we follow the design

TABLE 8.8: Simulation results generated by the BOIN12 shiny app. The values corresponding to OBD are in boldface.

	Dose level					Avg. N	Stop %
	1	2	3	4	5		
Scenario 1							
Pr(DLT)	0.03	**0.17**	0.35	0.50	0.65		
Pr(Efficacy)	0.12	**0.39**	0.43	0.55	0.65		
Mean utility	46.0	**56.6**	51.8	53.0	53.0		
No. patients treated	5.2	**8.8**	6.4	2.8	0.6	23.8	
Selection %	16.5	**53.5**	24.0	5.5	0.0		0.5
Scenario 2							
Pr(DLT)	0.01	0.08	0.10	**0.12**	0.35		
Pr(Efficacy)	0.05	0.25	0.30	**0.60**	0.60		
Mean utility	42.6	51.8	54.0	**71.2**	62.0		
No. patients treated	3.4	4.8	4.7	**9.0**	4.2	26.1	
Selection %	1.5	14.5	6.5	**63.5**	14.0		0.0

decision table included in the protocol to conduct the trial and make adaptive decisions (e.g., dose escalation/stay/de-escalation). Alternatively, users can use the Trial Conduct tab to determine the dose for next cohort of patients. In the latter approach, users can upload trial data to the app to obtain the recommended dose for the next cohort of patients. Summary statistics for the interim data are also provided by the app.

After the trial completes accrual and has all patients' outcomes evaluated, users can use the OBD determination tab to identify OBD. After uploading the trial data with the provided csv template, users can obtain OBD on the

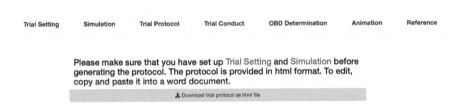

FIGURE 8.16: Download protocol templates of the BOIN12 design.

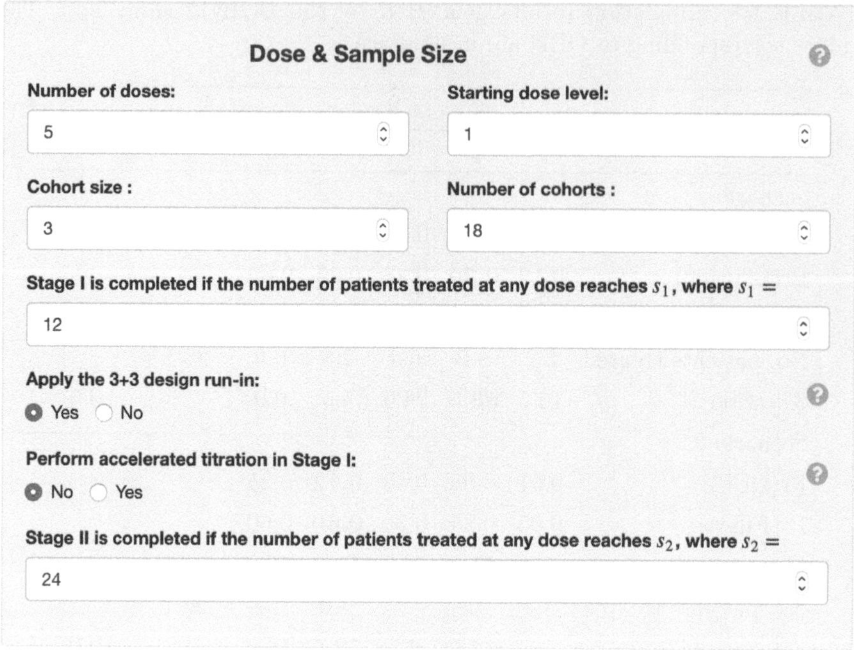

Dose & Sample Size   ❓

Number of doses:

5   ↕

Starting dose level:

1   ↕

Cohort size :

3   ↕

Number of cohorts :

18   ↕

Stage I is completed if the number of patients treated at any dose reaches s_1, where $s_1 =$

12   ↕

Apply the 3+3 design run-in: ❓
● Yes ○ No

Perform accelerated titration in Stage I: ❓
● No ○ Yes

Stage II is completed if the number of patients treated at any dose reaches s_2, where $s_2 =$

24   ↕

FIGURE 8.17: Dose and sample size of the U-BOIN design.

right side of the app, as well as various estimates, including the estimates for joint toxicity-efficacy probabilities, marginal toxicity/efficacy probability, and utility of each dose that has been used to treat patients.

Software for U-BOIN and TITE-BOIN12

The software for U-BOIN and TITE-BOIN12 has the same layout as BOIN12. The specification of risk-benefit trade-off and admissible criteria follow the same principle as described above. For U-BOIN, a few extra design parameters need to be specified, which are mostly related to the first stage of the design that focuses on the dose exploration based on toxicity using the standard BOIN design. As shown in Figure 8.17, the design parameters are similar to those of BOIN, e.g., apply the 3+3 design run-in and accelerated titration. Section 3.6 provides detailed discussion on how to choose these design parameters. For the Stage I "convergence" stopping rule, we recommend $m_1 \geq 9$. In U-BOIN, patients treated in Stage I roll over to Stage II, thus the Stage II "convergence" early stopping cutoff m_2 should be greater than m_1. Roughly speaking, $m_2 - m_1$ represents the maximum number of additional patients we may expand on each of admissible doses identified at the end of Stage I to optimize the dose. For example, in Figure 8.17, $m_2 = 24$ means treating up to 12 additional patients at OBD and/or admissible doses. The exact meaning of

$m_2 - m_1$ depends on which strategy we use to assign patients in Stage II (e.g., pick the winner, adaptive randomization, or equal randomization). Simulation should be performed to validate and calibrate the values of m_1 and m_2 to obtain desirable operating characteristics. As described previously, identification of OBD (or optimizing the dose in general) is substantially more challenging than identification of MTD. Using an excessively small m_1 and m_2 leads to very low power to identify OBD.

Bibliography

Ahn, C. (1998). An evaluation of phase I cancer clinical trial designs. *Statistics in Medicine*, 17(14):1537–1549.

Babb, J., Rogatko, A., and Zacks, S. (1998). Cancer phase I clinical trials: efficient dose escalation with overdose control. *Statistics in Medicine*, 17(10):1103–1120.

Barlow, R. E., Bartholomew, D. J., Bremner, J. M., and Brunk, H. D. (1972). *Statistical Inference under Order Restrictions; The Theory and Application of Isotonic Regression*. Wiley, New York, NY.

Bekele, B. N., Ji, Y., Shen, Y., and Thall, P. F. (2007). Monitoring late-onset toxicities in phase I trials using predicted risks. *Biostatistics*, 9(3):442–457.

Bekele, B. N. and Shen, Y. (2005). A Bayesian approach to jointly modeling toxicity and biomarker expression in a phase I/II dose-finding trial. *Biometrics*, 61(2):343–354.

Bekele, B. N. and Thall, P. F. (2004). Dose-finding based on multiple toxicities in a soft tissue sarcoma trial. *Journal of the American Statistical Association*, 99(465):26–35.

Berger, J. O. (2013). *Statistical Decision Theory and Bayesian Analysis*. Springer Science & Business Media.

Berry, D. A. (2003). Statistical innovations in cancer research. *Cancer Medicine*, 6:465–478.

Berry, S. M., Carlin, B. P., Lee, J. J., and Muller, P. (2010). *Bayesian Adaptive Methods for Clinical Trials*. CRC press.

Biswas, S., Liu, D. D., Lee, J. J., and Berry, D. A. (2009). Bayesian clinical trials at the University of Texas M. D. Anderson Cancer Center. *Clinical Trials*, 6(3):205–216.

Brahmer, J. R., Drake, C. G., Wollner, I., Powderly, J. D., Picus, J., Sharfman, W. H., Stankevich, E., Pons, A., Salay, T. M., McMiller, T. L., et al. (2010). Phase I study of single-agent anti–programmed death-1 (MDX-1106) in refractory solid tumors: safety, clinical activity, pharmacodynamics, and immunologic correlates. *Journal of Clinical Oncology*, 28:3167–3175.

Braun, T. M. and Jia, N. (2013). A generalized continual reassessment method for two-agent phase I trials. *Statistics in Biopharmaceutical Research*, 5(2):105–115.

Braun, T. M. and Wang, S. (2010). A hierarchical Bayesian design for phase I trials of novel combinations of cancer therapeutic agents. *Biometrics*, 66(3):805–812.

Bril, G., Dykstra, R., Pillers, C., and Robertson, T. (1984a). Algorithm AS 206: isotonic regression in two independent variables. *Journal of the Royal Statistical Society. Series C (Applied Statistics)*, 33(3):352–357.

Bril, G., Dykstra, R., Pillers, C., and Robertson, T. (1984b). Isotonic regression in two independent variables. *Journal of the Royal Statistical Society: Series C (Applied Statistics)*, 33:352–358.

Bugano, D. D., Hess, K., Jardim, D. L., Zer, A., Meric-Bernstam, F., Siu, L. L., Razak, A. R., and Hong, D. S. (2017). Use of expansion cohorts in phase I trials and probability of success in phase II for 381 anticancer drugs. *Clinical Cancer Research*, 23(15):4020–4026.

Cai, C., Yuan, Y., and Ji, Y. (2014). A Bayesian dose finding design for oncology clinical trials of combinational biological agents. *Journal of the Royal Statistical Society: Series C (Applied Statistics)*, 63(1):159–173.

Carlin, B. P. and Louis, T. A. (2008). *Bayesian Methods for Data Analysis*. CRC Press.

Chen, Z., Krailo, M., Azen, S., and Tighiouart, M. (2010). A novel toxicity scoring system treating toxicity response as a quasi-continuous variable in phase I clinical trials. *Contemporary Clinical Trials*, 31:473–482.

Cheung, Y. K. (2005). Coherence principles in dose-finding studies. *Biometrika*, 92(4):863–873.

Cheung, Y. K. (2011). *Dose Finding by the Continual Reassessment Method*. Chapman and Hall/CRC, Boca Raton, FL.

Cheung, Y. K. and Chappell, R. (2000). Sequential designs for phase I clinical trials with late-onset toxicities. *Biometrics*, 56(4):1177–1182.

Chow, S.-C. and Chang, M. (2008). Adaptive design methods in clinical trials–a review. *Orphanet Journal of Rare Diseases*, 3(1):1–13.

Clertant, M. and O'Quigley, J. (2017). Semiparametric dose finding methods. *Journal of the Royal Statistical Society Series B*, 79(5):1487–1508.

Cook, N., Hansen, A. R., Siu, L. L., and Razak, A. R. A. (2015). Early phase clinical trials to identify optimal dosing and safety. *Molecular Oncology*, 9(5):997–1007.

Dale, J. R. (1986). Global cross-ratio models for bivariate, discrete, ordered responses. *Biometrics*, 42(4):909–917.

Durham, S. D., Flournoy, N., and Rosenberger, W. F. (1997). A random walk rule for phase I clinical trials. *Biometrics*, 53:745–760.

Ezzalfani, M., Zohar, S., Qin, R., Mandrekar, S. J., and Deley, M.-C. L. (2013). Dose-finding designs using a novel quasi-continuous endpoint for multiple toxicities. *Statistics in Medicine*, 32:2728–2746.

Faries, D. (1994). Practical modifications of the continual reassessment method for phase I cancer clinical trials. *Journal of Biopharmaceutical Statistics*, 4(2):147–164.

FDA (2006). Nonclinical safety evaluation of drug or biologic combinations. *FDA Guidance for Industry*.

FDA (2013). Codevelopment of two or more new investigational drugs for use in combination. *FDA Guidance for Industry*.

FDA (2019). Submitting documents using real-world data and real-world evidence to FDA for drugs and biologics guidance for industry. *FDA Guidance for Industry*.

FDA (2022). Project Optimus, https://www.fda.gov/about-fda/oncology-center-excellence/project-optimus

Garrett-Mayer, E. (2006). The continual reassessment method for dose-finding studies: a tutorial. *Clinical Trials*, 3(1):57–71.

Gelman, A., Carlin, J. B., Stern, H. S., Dunson, D. B., Vehtari, A., and Rubin, D. B. (2013). *Bayesian Data Analysis*. Chapman and Hall/CRC.

Goodman, S., Zahurak, M., and Piantadosi, S. (1995). Some pratical improvements in the continual reassessment method for phase I studies. *Statistics in Medicine*, 14:1149–1161.

Guo, B. and Li, Y. (2015). Bayesian dose-finding designs for combination of molecularly targeted agents assuming partial stochastic ordering. *Statistics in Medicine*, 34(5):859–875.

Guo, B., Li, Y., and Yuan, Y. (2016). A dose–schedule finding design for phase I–II clinical trials. *Journal of the Royal Statistical Society: Series C (Applied Statistics)*, 65(2):259–272.

Guo, B. and Yuan, Y. (2017). Bayesian phase I/II biomarker-based dose finding for precision medicine with molecularly targeted agents. *Journal of the American Statistical Association*, 112(518):508–520.

Heyd, J. M. and Carlin, B. P. (1999). Adaptive design improvements in the continual reassessment method for phase I studies. *Statistics in Medicine*, 18(11):1307–1321.

Houede, N., Thall, P. F., Nguyen, H., Paoletti, X., and Kramar, A. (2010). Utility-based optimization of combination therapy using ordinal toxicity and efficacy in phase I/II trials. *Biometrics*, 66(2):532–540.

Hunsberger, S., Rubinstein, L. V., Dancey, J., and Korn, E. L. (2005). Dose escalation trial designs based on a molecularly targeted endpoint. *Statistics in Medicine*, 24(14):2171–2181.

Iasonos, A., Wages, N. A., Conaway, M. R., Cheung, K., Yuan, Y., and O'Quigley, J. (2016). Dimension of model parameter space and operating characteristics in adaptive dose-finding studies. *Statistics in Medicine*, 35(21):3760–3775.

Iasonos, A., Wilton, A. S., Riedel, E. R., Seshan, V. E., and Spriggs, D. R. (2008). A comprehensive comparison of the continual reassessment method to the standard 3+3 dose escalation scheme in phase I dose-finding studies. *Clinical Trials*, 5(5):465–477.

Ivanova, A., Montazer-Haghighi, A., Mohanty, S. G., and Durham, S. D. (2003). Improved up-and-down designs for phase I trials. *Statistics in Medicine*, 22(1):69–82.

Jaki, T., Clive, S., and Weir, C. J. (2013). Principles of dose finding studies in cancer: a comparison of trial designs. *Cancer Chemotherapy and Pharmacology*, 71(5):1107–1114.

Ji, Y., Liu, P., Li, Y., and Nebiyou Bekele, B. (2010). A modified toxicity probability interval method for dose-finding trials. *Clinical Trials*, 7(6):653–663.

Kass, R. E. and Raftery, A. E. (1995). Bayes factors. *Journal of the American Statistical Association*, 90(430):773–795.

Le Tourneau, C., Diéras, V., Tresca, P., Cacheux, W., and Paoletti, X. (2010). Current challenges for the early clinical development of anticancer drugs in the era of molecularly targeted agents. *Targeted Oncology*, 5:65–72.

Le Tourneau, C., Lee, J. J., and Siu, L. L. (2009). Dose escalation methods in phase I cancer clinical trials. *JNCI: Journal of the National Cancer Institute*, 101(10):708–720.

Lee, J., Thall, P. F., Ji, Y., and Müller, P. (2016). A decision-theoretic phase I–II design for ordinal outcomes in two cycles. *Biostatistics*, 17(2):304–319.

Lee, J. J. and Chu, C. T. (2012). Bayesian clinical trials in action. *Statistics in Medicine*, 31(25):2955–2972.

Lee, S., Hershman, D., Martin, P., Leonard, J., and Cheung, K. (2009). Validation of toxicity burden score for use in phase I clinical trials. *Journal of Clinical Oncology*, 27(15_suppl):2514–2514.

Lee, S. M., Cheng, B., and Cheung, Y. K. (2010). Continual reassessment method with multiple toxicity constraints. *Biostatistics*, 12(2):386–398.

Lee, S. M. and Cheung, Y. K. (2009). Model calibration in the continual reassessment method. *Clinical Trials*, 6(3):227–238.

Lee, S. M., Ursino, M., Cheung, Y. K., and Zohar, S. (2017). Dose-finding designs for cumulative toxicities using multiple constraints. *Biostatistics*, 20(1):17–29.

Li, D. H., Whitmore, J. B., Guo, W., and Ji, Y. (2017). Toxicity and efficacy probability interval design for phase I adoptive cell therapy dose-finding clinical trials. *Clinical Cancer Research*, 23(1):13–20.

Li, Y. and Yuan, Y. (2020). PA-CRM: a continuous reassessment method for pediatric phase I oncology trials with concurrent adult trials. *Biometrics*, 76(4):1364–1373.

Lin, R. (2018). Bayesian optimal interval design with multiple toxicity constraints. *Biometrics*, 74(4):1320–1330.

Lin, R. and Lee, J. J. (2020). Novel Bayesian adaptive designs and their applications in cancer clinical trials. In *Computational and Methodological Statistics and Biostatistics*, 395–426. Springer.

Lin, R., Thall, P. F., and Yuan, Y. (2020a). An adaptive trial design to optimize dose-schedule regimes with delayed outcomes. *Biometrics*, 76(1):304–315.

Lin, R., Thall, P. F., and Yuan, Y. (2021). A phase I–II basket trial design to optimize dose-schedule regimes based on delayed outcomes. *Bayesian Analysis*, 16(1):179–202.

Lin, R. and Yin, G. (2017a). Bayesian optimal interval design for dose finding in drug-combination trials. *Statistical Methods in Medical Research*, 26(5):2155–2167.

Lin, R. and Yin, G. (2017b). STEIN: a simple toxicity and efficacy interval design for seamless phase I/II clinical trials. *Statistics in Medicine*, 36(26):4106–4120.

Lin, R. and Yuan, Y. (2019). On the relative efficiency of model-assisted designs: a conditional approach. *Journal of Biopharmaceutical Statistics*, 29(4):648–662.

Lin, R. and Yuan, Y. (2020). Time-to-event model-assisted designs for dose-finding trials with delayed toxicity. *Biostatistics*, 21(4):807–824.

Lin, R., Zhou, Y., Yan, F., Li, D., and Yuan, Y. (2020b). BOIN12: Bayesian optimal interval phase I/II trial design for utility-based dose finding in immunotherapy and targeted therapies. *JCO Precision Oncology*, 4:1393–1402.

Lin, Y. and Shih, W. J. (2001). Statistical properties of the traditional algorithm-based designs for phase I cancer clinical trials. *Biostatistics*, 2(2):203–215.

Little, R. J. and Rubin, D. B. (2014). *Statistical Analysis with Missing Data*, volume 333. John Wiley & Sons.

Liu, R., Yuan, Y., Sen, S., Yang, X., Jiang, Q., Li, X., Lu, C., Gonen, M., Tian, H., Zhou, H., Lin, R., and Marchenko, O. (2022). Accuracy and safety of novel designs for phase I drug-combination oncology trials. *Statistics in Biopharmaceutical Research*, 14(3):270–282.

Liu, S., Guo, B., and Yuan, Y. (2018). A Bayesian phase I/II trial design for immunotherapy. *Journal of the American Statistical Association*, 113(523):1016–1027.

Liu, S. and Johnson, V. E. (2016). A robust Bayesian dose-finding design for phase I/II clinical trials. *Biostatistics*, 17(2):249–263.

Liu, S., Pan, H., Xia, J., Huang, Q., and Yuan, Y. (2015). Bridging continual reassessment method for phase I clinical trials in different ethnic populations. *Statistics in Medicine*, 34(10):1681–1694.

Liu, S., Yin, G., and Yuan, Y. (2013). Bayesian data augmentation dose finding with continual reassessment method and delayed toxicity. *The Annals of Applied Statistics*, 7(4):1837.

Liu, S. and Yuan, Y. (2015). Bayesian optimal interval designs for phase I clinical trials. *Journal of the Royal Statistical Society: Series C (Applied Statistics)*, 64(3):507–523.

Liu, S. and Yuan, Y. (2022). Bayesian optimal interval designs for phase I clinical trials. *Journal of the Royal Statistical Society: Series C (Applied Statistics)*, 71(2):491–492.

Mahajan, R. and Gupta, K. (2010). Adaptive design clinical trials: methodology, challenges and prospect. *Indian Journal of Pharmacology*, 42(4):201.

Mander, A. P. and Sweeting, M. J. (2015). A product of independent beta probabilities dose escalation design for dual-agent phase I trials. *Statistics in Medicine*, 34(8):1261–1276.

Mandrekar, S. J., Qin, R., and Sargent, D. J. (2010). Model-based phase I designs incorporating toxicity and efficacy for single and dual agent drug combinations: methods and challenges. *Statistics in Medicine*, 29(10):1077–1083.

Mathijssen, R. H., Sparreboom, A., and Verweij, J. (2014). Determining the optimal dose in the development of anticancer agents. *Nature Reviews Clinical Oncology*, 11(5):272.

Morita S, Thall PF, Müller P. (2008) Determining the effective sample size of a parametric prior. *Biometrics*, 64(2):595–602.

Morita, S. (2011). Application of the continual reassessment method to a phase I dose-finding trial in Japanese patients: east meets west. *Statistics in Medicine*, 30(17):2090–2097.

Mu, R., Yuan, Y., Xu, J., Mandrekar, S. J., and Yin, J. (2018). gBOIN: a unified model-assisted phase I trial design accounting for toxicity grades, and binary or continuous end points. *Journal of the Royal Statistical Society: Series C*, 68(2):289–308.

Murray, T. A., Yuan, Y., Thall, P. F., Elizondo, J. H., and Hofstetter, W. L. (2018). A utility-based design for randomized comparative trials with ordinal outcomes and prognostic subgroups, *Biometrics*, 74(3):1095–1103.

Neuenschwander, B., Branson, M., and Gsponer, T. (2008). Critical aspects of the Bayesian approach to phase I cancer trials. *Statistics in Medicine*, 27(13):2420–2439.

Neuenschwander, B., Matano, A., Tang, Z., Roychoudhury, S., Wandel, S., and Bailey, S. (2015). Bayesian industry approach to phase I combination trials in oncology. *Statistical Methods in Drug Combination Studies*, 6:95–135.

Onar-Thomas, A. and Xiong, Z. (2010). A simulation-based comparison of the traditional method, Rolling-6 design and a frequentist version of the continual reassessment method with special attention to trial duration in pediatric phase I oncology trials. *Contemporary Clinical Trials*, 31(3):259–270.

O'Quigley, J. and Chevret, S. (1991). Methods for dose finding studies in cancer clinical trials: a review and results of a monte carlo study. *Statistics in Medicine*, 10(11):1647–1664.

O'Quigley, J., Pepe, M., and Fisher, L. (1990). Continual reassessment method: a practical design for phase 1 clinical trials in cancer. *Biometrics*, 46(1):33–48.

Pallmann, P., Bedding, A. W., Choodari-Oskooei, B., Dimairo, M., Flight, L., Hampson, L. V., Holmes, J., Mander, A. P., Sydes, M. R., Villar, S. S., et al. (2018). Adaptive designs in clinical trials: why use them, and how to run and report them. *BMC Medicine*, 16(1):1–15.

Pan, H., Lin, R., Zhou, Y., and Yuan, Y. (2020). Keyboard design for phase I drug-combination trials. *Contemporary Clinical Trials*, 92:105972.

Pan, H. and Yuan, Y. (2017). A default method to specify skeletons for Bayesian model averaging continual reassessment method for phase I clinical trials. *Statistics in Medicine*, 36(2):266–279.

Papke, L. E. and Wooldridge, J. M. (1996). Econometric methods for fractional response variables with an application to 401 (k) plan participation rates. *Journal of Applied Econometrics*, 11(6):619–632.

Penel, N., Adenis, A., Clisant, S., and Bonneterre, J. (2011). Nature and subjectivity of dose-limiting toxicities in contemporary phase I trials: comparison of cytotoxic versus non-cytotoxic drugs. *Investigational New Drugs*, 29:1414–1419.

Petit, C., Samson, A., Morita, S., Ursino, M., Guedj, J., Jullien, V., Comets, E., and Zohar, S. (2018). Unified approach for extrapolation and bridging of adult information in early-phase dose-finding paediatric studies. *Statistical Methods in Medical Research*, 27(6):1860–1877.

Phan, T. G., Ma, H., Lim, R., Sobey, C. G., and Wallace, E. M. (2018). Phase 1 trial of amnion cell therapy for ischemic stroke. *Frontiers in Neurology*, 9:198.

Piantadosi, S., Fisher, J. D., and Grossman, S. (1998). Practical implementation of a modified continual reassessment method for dose-finding trials. *Cancer Chemotherapy and Pharmacology*, 41(6):429–436.

Postel-Vinay, S., Gomez-Roca, C., Molife, L. R., Anghan, B., Levy, A., Judson, I., De Bono, J., Soria, J.-C., Kaye, S., and Paoletti, X. (2011). Phase I trials of molecularly targeted agents: should we pay more attention to late toxicities. *Journal of Clinical Oncology*, 29(13):1728–1735.

Riviere, M.-K., Yuan, Y., Dubois, F., and Zohar, S. (2014). A Bayesian dose-finding design for drug combination clinical trials based on the logistic model. *Pharmaceutical Statistics*, 13(4):247–257.

Riviere, M.-K., Dubois, F., and Zohar, S. (2015a). Competing designs for drug combination in phase I dose-finding clinical trials. *Statistics in Medicine*, 34(1):1–12.

Riviere, M.-K., Yuan, Y., Dubois, F., and Zohar, S. (2015b). A Bayesian dose finding design for clinical trials combining a cytotoxic agent with a molecularly targeted agent. *Journal of the Royal Statistical Society: Series C (Applied Statistics)*, 64(1):215–229.

Riviere, M.-K., Yuan, Y., Jourdan, J.-H., Dubois, F., and Zohar, S. (2018). Phase I/II dose-finding design for molecularly targeted agent: plateau determination using adaptive randomization. *Statistical Methods in Medical Research*, 27(2):466–479.

Robert, C. and Casella, G. (2013). *Monte Carlo Statistical Methods*. Springer Science & Business Media.

Rogatko, A., Schoeneck, D., Jonas, W., Tighiouart, M., Khuri, F. R., and Porter, A. (2007). Translation of innovative designs into phase I trials. *Journal of Clinical Oncology*, 25(31):4982–4986.

Ruppert, A. S. and Shoben, A. B. (2018). Overall success rate of a safe and efficacious drug: results using six phase 1 designs, each followed by standard phase 2 and 3 designs. *Contemporary Clinical Trials Communications*, 12:40–50.

Sachs, J. R., Mayawala, K., Gadamsetty, S., Kang, S. P., and de Alwis, D. P. (2016). Optimal dosing for targeted therapies in oncology: drug development cases leading by example. *Clinical Cancer Research*, 22(6):1318–1324.

Shah, M., Rahman, A., Theoret, M., Pazdur, R. (2021). The Drug-Dosing Conundrum in Oncology - When Less Is More. *The New England Journal of Medicine*, 385:1445–1447.

Shen, L. Z. and O'Quigley, J. (1996). Consistency of continual reassessment method under model misspecification. *Biometrika*, 83(2):395.

Shi, H., Cao, J., Yuan, Y., and Lin, R. (2021). uTPI: a utility-based toxicity probability interval design for phase I/II dose-finding trials. *Statistics in Medicine*, 40(11):2626–2649.

Simon, R. (1989). Optimal two-stage designs for phase II clinical trials. *Controlled Clinical Trials*, 10(1):1–10.

Simon, R., Freidlin, B., Rubinstein, L., Arbuck, S. G., Collins, J., and Christian, M. C. (1997). Accelerated titration designs for phase I clinical trials in oncology. *Journal of the National Cancer Institute*, 89(15):1138–1147.

Skolnik, J. M., Barrett, J. S., Jayaraman, B., Patel, D., and Adamson, P. C. (2008). Shortening the timeline of pediatric phase I trials: the rolling six design. *Journal of Clinical Oncology*, 26(2):190–195.

Storer, B. E. (1989). Design and analysis of phase I clinical trials. *Biometrics*, 45(3):925–937.

Stylianou, M. and Follmann, D. A. (2004). The accelerated biased coin up-and-down design in phase I trials. *Journal of Biopharmaceutical Statistics*, 14(1):249–260.

Takeda, K., Taguri, M., and Morita, S. (2018). BOIN-ET: Bayesian optimal interval design for dose finding based on both efficacy and toxicity outcomes. *Pharmaceutical Statistics*, 17(4):383–395.

Takeda K., Morita S., and Taguri M. (2020). TITE-BOIN-ET: Time-to-event Bayesian optimal interval design to accelerate dose-finding based on both efficacy and toxicity outcomes. *Pharmaceutical Statistics*, 19(3):335–349.

Thall, P. F. and Cook, J. D. (2004). Dose-finding based on efficacy–toxicity trade-offs. *Biometrics*, 60(3):684–693.

Thall, P. F., Millikan, R. E., Mueller, P., and Lee, S.-J. (2003). Dose-finding with two agents in phase I oncology trials. *Biometrics*, 59(3):487–496.

Tidwell, R. S. S., Peng, S. A., Chen, M., Liu, D. D., Yuan, Y., and Lee, J. J. (2019). Bayesian clinical trials at the University of Texas MD Anderson Cancer Center: an update. *Clinical Trials*, 16(6):645–656.

van Brummelen, E. M., Huitema, A. D., van Werkhoven, E., Beijnen, J. H., and Schellens, J. H. (2016). The performance of model-based versus rule-based phase I clinical trials in oncology. *Journal of Pharmacokinetics and Pharmacodynamics*, 43(3):235–242.

Wages, N. A., Conaway, M. R., and O'Quigley, J. (2011). Dose-finding design for multi-drug combinations. *Clinical Trials*, 8(4):380–389.

Weber, J. S., Yang, J. C., Atkins, M. B., and Disis, M. L. (2015). Toxicities of immunotherapy for the practitioner. *Journal of Clinical Oncology*, 33(18):2092.

Yan, F., Mandrekar, S. J., and Yuan, Y. (2017). Keyboard: a novel Bayesian toxicity probability interval design for phase I clinical trials. *Clinical Cancer Research*, 23(15):3994–4003.

Yin, G., Li, Y., and Ji, Y. (2006). Bayesian dose-finding in phase I/II clinical trials using toxicity and efficacy odds ratios. *Biometrics*, 62(3):777–787.

Yin, G. and Lin, R. (2015). Comments on 'competing designs for drug combination in phase I dose-finding clinical trials' by M-K. Riviere, F. Dubois, and S. Zohar. *Statistics in Medicine*, 34(1):13–17.

Yin, G. and Yuan, Y. (2009a). Bayesian dose finding in oncology for drug combinations by copula regression. *Journal of the Royal Statistical Society: Series C (Applied Statistics)*, 58(2):211–224.

Yin, G. and Yuan, Y. (2009b). Bayesian model averaging continual reassessment method in phase I clinical trials. *Journal of the American Statistical Association*, 104(487):954–968.

Yin, G. and Yuan, Y. (2009c). A latent contingency table approach to dose finding for combinations of two agents. *Biometrics*, 65(3):866–875.

Yin, G., Zheng, S., and Xu, J. (2013). Fractional dose-finding methods with late-onset toxicity in phase I clinical trials. *Journal of Biopharmaceutical Statistics*, 23(4):856–870.

Yuan, Y., Hess, K. R., Hilsenbeck, S. G., and Gilbert, M. R. (2016a). Bayesian optimal interval design: a simple and well-performing design for phase I oncology trials. *Clinical Cancer Research*, 22(17):4291–4301.

Yuan, Y., Lee, J. J., and Hilsenbeck, S. G. (2019). Model-assisted designs for early-phase clinical trials: simplicity meets superiority. *JCO Precision Oncology*, 3:1–12.

Yuan, Y., Lin, R., Li, D., Nie, L., and Warren, K. E. (2018). Time-to-event Bayesian optimal interval design to accelerate phase I trials. *Clinical Cancer Research*, 24(20):4921–4930.

Yuan, Y., Nguyen, H. Q., and Thall, P. F. (2016b). *Bayesian Designs for Phase I–II Clinical Trials*. CRC Press.

Yuan, Y. and Yin, G. (2011a). Bayesian phase I/II adaptively randomized oncology trials with combined drugs. *The Annals of Applied Statistics*, 5(2A):924.

Yuan, Y. and Yin, G. (2011b). Robust EM continual reassessment method in oncology dose finding. *Journal of the American Statistical Association*, 106(495):818–831.

Yuan, Z., Chappell, R., and Bailey, H. (2007). The continual reassessment method for multiple toxicity grades: a Bayesian quasi-likelihood approach. *Biometrics*, 63(1):173–179.

Zang, Y. and Lee, J. J. (2014). Adaptive clinical trial designs in oncology. *Chinese Clinical Oncology*, 3(4):49.

Zang, Y., Lee, J. J., and Yuan, Y. (2014). Adaptive designs for identifying optimal biological dose for molecularly targeted agents. *Clinical Trials*, 11(3):319–327.

Zhang, L. and Yuan, Y. (2016). A practical Bayesian design to identify the maximum tolerated dose contour for drug combination trials. *Statistics in Medicine*, 35(27):4924–4936.

Zhang, W., Sargent, D. J., and Mandrekar, S. (2006). An adaptive dose-finding design incorporating both toxicity and efficacy. *Statistics in Medicine*, 25(14):2365–2383.

Zhao, L., Lee, J., Mody, R., and Braun, T. M. (2011). The superiority of the time-to-event continual reassessment method to the rolling six design in pediatric oncology phase I trials. *Clinical Trials*, 8(4):361–369.

Zhou, H., Murray, T. A., Pan, H., and Yuan, Y. (2018a). Comparative review of novel model-assisted designs for phase I clinical trials. *Statistics in Medicine*, 37(14):2208–2222.

Zhou, H., Yuan, Y., and Nie, L. (2018b). Accuracy, safety, and reliability of novel phase I trial designs. *Clinical Cancer Research*, 24(18):4357–4364.

Zhou, Y., Lee, J. J., Wang, S., Bailey, S., and Yuan, Y. (2021a). Incorporating historical information to improve phase I clinical trials. *Pharmaceutical Statistics*, 20(6):1017–1034.

Zhou, Y., Lee, J. J., and Yuan, Y. (2019). A utility-based Bayesian optimal interval (U-BOIN) phase I/II design to identify the optimal biological dose for targeted and immune therapies. *Statistics in Medicine*, 38(28):S5299–S5316.

Zhou, Y., Li, R., Yan, F., Lee, J. J., and Yuan, Y. (2021b). A comparative study of Bayesian optimal interval (BOIN) design with interval 3 + 3 (i3+3) design for phase I oncology dose-finding trials. *Statistics in Biopharmaceutical Research*, 13(2):147–155.

Zhou, Y., Lin, R., Lee, J. J., Li, D., Wang, L., Li, R., and Yuan, Y. (2022). TITE-BOIN12: a Bayesian phase I/II trial design to find the optimal biological dose with late-onset toxicity and efficacy. *Statistics in Medicine*, 41(11):1918–1931.

Zohar, S., Katsahian, S., and O'Quigley, J. (2011). An approach to meta-analysis of dose-finding studies. *Statistics in Medicine*, 30(17):2109–2116.

Index